# Electrical and Electronic Measurement and Testing

# ELECTRICAL AND ELECTRONIC MEASUREMENT AND TESTING

## W Bolton

**Longman Scientific & Technical**

**Longman Scientific & Technical**
Longman Group UK Ltd,
Longman House, Burnt Mill, Harlow,
Essex CM20 2JE, England
*and Associated Companies throughout the world.*

First published 1992

ISBN 0582 089670

**British Library Cataloguing in Publication Data**
A catalogue record for this book is available from the British Library


Set by 8 in 10/12pt Times

Printed in Malaysia

# Contents

# Preface

This book is an overview of the instruments and methods of electrical and electronic measurement and testing with the aims of introducing the reader to:

1 the concepts of measurement and testing, including sources of errors, reliability and standards;
2 the terms used in performance specifications;
3 the principles of data processing and display;
4 the principles and characteristics of basic and specialist test equipment – analogue meters, digital meters, recorders, bridges, wattmeters, oscilloscopes, counters, signal sources, digital circuit test instruments and automatic instruments;
5 the principles of measurement systems – transducers, signal conditioning and processing and data transmission;
6 the principles of test procedures for manual and automatic testing.

Chapters 1 to 5 are a consideration of the general concepts of measurement and testing and the terminology used in specifications of the performance of instruments, Chapters 6 to 16 the principles of electrical and electronic instruments and their typical performance characteristics, Chapter 17 the principles of measurement systems, and Chapter 18 the principles of test procedures. A knowledge of basic electrical principles has been assumed. The discussions of the principles of basic test instruments includes a brief consideration of their basic physical principles, with a systems approach being adopted for more complex instruments. The aim throughout has been to give sufficient information to enable the reader to be able to make an intelligent selection of instrument for a specific purpose and interpret the specifications given for instruments.

The book is seen as being relevant to all electrical and electronic technicians and engineers requiring an overview of measurement and testing instruments and methods. It covers the range of instrumentation and methods included in BTEC courses for electrical and electronic technicians at National Certificate/Diploma and Higher National Certificate/Diploma and, in particular, more than covers the BTEC unit Measurement and Testing H13682B.

W. Bolton

# 1 Measurement and testing

**Introduction**

The term *measurement* is used to describe the act of determining the value or size of some quantity, e.g., a current. The term *testing* is used if the measurements are taken to ascertain whether some product conforms to specified standards and quality. This chapter is a consideration of the types of electrical and electronic measurements and testing commonly carried out, sources of error that can arise, and the statistical analysis of errors. As such it is a general introduction to measurement and testing with details of specific electrical and electronic measurement techniques and instruments following in later chapters.

**Methods of measurement**

There are a number of ways methods of measurement can be classified. One useful way with electrical and electronic measurement methods is a grouping into the following three categories:

1   *Analogue measurements*   The quantity being measured is continuously monitored and the instrument used gives a response which is an analogue of the quantity, i.e., the magnitude of the instrument output represents the size of the quantity being measured. A moving coil galvanometer is an example of such an instrument, the size of the angular deflection of the meter pointer being related to the size of the current flowing into the meter.

2   *Comparison measurements*   The quantity being measured is compared with standards and its value given when a match is obtained. A simple example of this is the *substitution method* of determining resistance. The current through the unknown resistance is compared with that through a standard resistance box. The resistance box is adjusted until the current is the same (Fig. 1.1). Another form of comparison method is the *null method* where the difference between the unknown and the known is

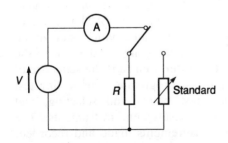

**Fig. 1.1**   Substitution method for resistance

detected and reduced to zero. The method is thus based on detecting the null condition. Examples of this are the potentiometer method of determining voltages and the Wheatstone bridge method of determining resistance.

3  *Digital measurements* With digital instruments the quantity being measured is sampled at regular intervals of time and the value of the sample converted into a number, i.e., a sequence of digits. The digital voltmeter is an example. Such an instrument samples the voltage, converts the sample into digits, waits a while and then repeats the process.

## Testing

Testing is carried out to ensure that a product conforms to its specification. The need to test can occur during the development of a product to confirm design decisions, during manufacture to control quality, at acceptance by the customer to ensure the product is up to specification, and during its operational life to ensure that it is continuing to perform to specification and to diagnose faults. In the case of an electrical system, testing means measuring the parameters, such as perhaps voltage at various points in the circuit, and characteristics, such as the output when the input is a square wave signal, and checking that they fall within those specified. To be able to carry out such testing it is necessary for the system under test to have a layout that permits testing to be carried out easily. The system is said to need to be designed for *testability*. A testable design is defined as being one that has built-in facilities that allow simple, efficient and effective testing to be carried out.

*Manual testing* involves a human carrying out the test procedure. This involves reading the test schedule and test instructions, assembling the various items of test equipment and connecting them to the unit under test, applying and adjusting test input signals, selecting appropriate ranges on instruments, recording the results, comparing the results with those given in the test specification, interpreting the meaning of the comparison, repeating the sequence for each of the tests required and then producing a test report. *Automatic testing*, by means of automatic testing equipment (ATE), involves connecting the unit under test to the automatic tester via an interface which connects the various points on the unit to the tester. The control unit in the tester then runs the sequence of tests in accordance with the test programme, connecting test signals to the appropriate test points and selecting, and connecting, the appropriate measurement instruments. The tests are carried out, the measurements made and recorded, and then the sequence automatically repeated for each of the tests required. The results of the testing are then automatically compared with the standard preprogrammed values, interpreted and the outcome indicated as perhaps pass or fail.

Automatic testing has the advantages of reducing the demand for skilled test operators, reducing the chance of human error, eliminating interpretative variations of individual test operators, increasing the speed of testing, permitting the use of more complex test procedures, offering the opportunity to process more detailed test data and the possibility of automatic self-testing. It has the disadvantages of requiring highly skilled test and programming engineers to plan and programme the test system, the need to identify all the test needs and possible faults prior to putting the system into operation, relative inflexibility to changes in test requirements, the need for a complex interface between the ATE and the unit under test, greater complexity than a manual test system and so possibly less reliable, and higher initial cost of the system. Though ATE has a higher initial cost than manual test equipment, the increased speed of testing and the reduction in labour costs can make the equipment more cost effective in the long run when there many units to be tested.

**Error**

Whatever the type of measurement there will be errors. The *error* of a measurement is the difference between the result of the measurement and the true value of the quantity being measured:

$$\text{error} = \text{measured value} - \text{true value} \qquad [1]$$

The error is positive if the measured value is greater than the true value and negative if it is less than the true value. The *percentage error* is the error as a percentage of the true value, i.e.,

$$\text{percentage error} = \frac{\text{error}}{\text{true value}} \times 100\% \qquad [2]$$

The *accuracy* of a measurement is the extent to which it differs from the true value, i.e., the degree of uncertainty. Accuracy is frequently quoted as a percentage of the true value, i.e.,

$$\text{accuracy} = \frac{\text{measured value} - \text{true value}}{\text{true value}} \times 100\% \qquad [3]$$

Thus an accuracy of $\pm 1\%$ means that the measured value will lie within + or $-$ 1% of the true value.

In the case of some components and instruments their deviations from the specified values are guaranteed to be within a certain percentage of that value. The deviations in this case are then referred to as *limiting errors* or *tolerances*. Thus a nominally 100 Ω resistor may be marked as having a tolerance of ±10%. This means that the resistor can have an actual value anywhere within 10% of 100 Ω, i.e., from 90 to 110 Ω. The limiting errors or tolerances give the worst possible error that can occur.

### Example 1

What is the error if a voltmeter gives a reading of 10.1 V when the true reading is 10.5 V?

*Answer*

The error is the difference between the measured and true values (equation [1]). Thus

$$\text{error} = 10.1 - 10.5 = -0.4\,\text{V}$$

## Sources of error

Errors in general can be classified as being either random or systematic errors, though there is a group of errors which may be described as human errors. *Random errors* are ones that vary in an unpredictable manner between successive readings of the same quantity, varying both in the size of the error and whether the error is positive or negative. *Systematic errors* are errors that remain constant with repeated measurements. *Human errors* are the mistakes made by humans in using instruments and taking the readings.

A measurement is said to be *accurate* when the random, systematic and human errors are small. Accuracy is how close the measured value agrees with the true value. A measurement is said to be *precise* when the random errors are small, regardless of whether or not the systematic and human errors are small. A precise result is one for which repeated measurements give only small variations. Because random errors often arise from operators having to read instrument scales and estimate values when a pointer is between scale divisions, precision is often indicated by how easy it is to estimate a scale reading. Thus an instrument which has more graduations than another is likely to be more precise.

The following are common sources of random error:

1   *Operating errors*   These are errors that arise because an operator is taking the measurement. They are not mistakes but errors due to situations that lead to small variations in the readings perceived by operators. They include the errors in reading the position of a pointer on a scale due to the scale and pointer not being in the same plane, the reading obtained then depending on the angle at which the pointer is viewed against the scale, the so-

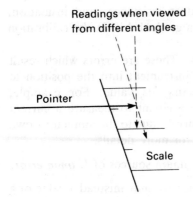

**Fig. 1.2**  Parallax error

called parallax error (Fig. 1.2). There are also the errors due to the uncertainty resulting from operators estimating readings between scale markings on an instrument's display or when such readings are rounded up or down to the nearest graduation.

2  *Environmental errors*  These are errors which can arise as a result of environmental effects, such as fluctuations in temperature, humidity or atmospheric pressure. Thus, for example, a change in temperature can produce a change in electrical resistance and thus change the resistance of the coil of a moving coil galvanometer and so affect its calibration.

3  *Stochastic errors*  The pressure due to a gas is inherently random since it results from the random motion of the gas molecules. Because of this the pressure will fluctuate in a random manner, and this will be detected by an instrument if it is sensitive enough. Another example of an inherently random process is electrical noise (see Chapter 3). The term stochastic process is used for processes which are inherently random and stochastic error for the random errors which result.

The following are common sources of *systematic error*:

1  *Construction errors*  These errors result from the manufacture of an instrument and the components used. They arise from such causes as tolerances on the dimensions of components and on the values of electrical components used.

2  *Equipment errors*  There may be a fault in an instrument which makes the calibration incorrect.

3  *Zero errors*  These arise from an incorrect setting of the instrument zero and will result in an instrument reading high or low over its entire range.

4  *Calibration errors*  An incorrect calibration may result, for example, in an instrument reading high or low over its entire range.

5  *Approximation errors*  These arise from assumptions made regarding relationships between quantities, e.g. a linear relationship between two quantities is often assumed and may in practice only be an approximation to the true relationship. Thus, for example, though an instrument may be shown as having a linear scale, i.e., equally spaced scale divisions, because the linear relationship is only approximately true the divisions should not be precisely equally spaced.

6  *Ageing errors*  These are errors resulting from instruments getting older, e.g., bearings wearing, components deteriorating and their values changing, a build-up of deposits

on surfaces affecting contact resistances and insulation. Such errors may show as a gradual drift in the calibration of an instrument.

7 *Insertion or loading errors* These are errors which result from the insertion of the instrument into the position to measure a quantity affecting its value. For example, inserting an ammeter into a circuit to measure a current changes the value of the current due to the ammeter's own resistance. See Chapter 3 for more details.

Finally, the following are common sources of *Human error*:

1 *Misreading errors* The operator may misread a value or a scale.
2 *Calculation errors* The operator may make a mistake in carrying out a calculation.
3 *Incorrect instrument* The operator may choose the wrong instrument or measurement method and so obtain incorrect results. For example, an instrument may be used to measure the voltage of a signal which is at a frequency greater than that for which the instrument is designed. The result would be a false reading.
4 *Incorrect adjustment* The operator may incorrectly adjust some aspect of the measurement system, e.g., incorrectly set the balance condition with a bridge or set the zero on a galvanometer.

**Estimation and reduction of errors**

Random errors can be estimated by taking many readings and applying statistical analysis (see later in this chapter). Such errors can be reduced by careful design of the measurement system to minimize interference or environmental fluctuations. This could involve using a constant temperature enclosure, mounting the instrument on a vibration isolation mount, and shielding against electric and magnetic fields. See Chapter 3 for a discussion of noise.

Systematic errors can be determined and allowed for. See Chapter 3 for a discussion of loading. Many systematic errors can be reduced by careful inspection and maintenance of instruments to ensure proper operation.

Human errors cannot be estimated. They can be reduced by using two or more operators to take readings or carry out operations but the main way of reducing such errors is by operators paying careful attention to what they are doing and fully understanding the techniques used and the limitations and capabilities of the instruments.

**The statistics of errors**

Random errors will lead to repeated measurements of a

particular quantity giving a spread of values. Statistical methods can be used to find the most probable value and indicate the probable error with any one measurement..

The most frequently used technique to arrive at the probable value of a set of measurements of some quantity subject to random errors is the determination of the *arithmetic mean*. This is the sum of all the results $x_1$, $x_2$, etc., divided by the number of results $n$ considered.

$$\text{arithmetic mean} = \bar{x} = \frac{x_1 + x_2 + x_3 + \ldots + x_n}{n} = \frac{\Sigma x}{n} \quad [4]$$

The larger the number of readings the more reliable is the mean and the more random errors are averaged out. With a very large number of results the mean may be regarded as completely reliable, with all random errors averaged out. Such a mean is sometimes called the *true mean* or *best value*.

The arithmetic mean does not give any indication of the spread of the results about the mean, i.e., the sizes of the random errors. Both the sets of results in Fig. 1.3 give the same arithmetic mean but one is spread more widely round the mean than the other. The mean value of a set of results is more likely to be close to the true mean if the spread of the results is small.

The term *deviation* is used for the amount by which an individual measurement departs from the mean. Thus for a measurement of $x_1$

$$\text{deviation} = x_1 - \bar{x} \quad [5]$$

with $\bar{x}$ being the mean value. The deviation from the mean may have a positive or a negative value. A measure of the spread of the results is then given by the *mean deviation*. This is the mean of the deviations, ignoring the signs of each deviation.

$$\text{mean deviation} = \frac{|x_1 - \bar{x}| + |x_2 - \bar{x}| + \ldots |x_n - \bar{x}|}{n}$$

$$= \frac{\Sigma |x_n - \bar{x}|}{n} \quad [6]$$

The smaller the mean deviation, the greater the precision of an instrument.

A more useful measure (see the next section for why it is useful) of the spread of results is the *standard deviation* or *root-mean-square deviation*. The deviation for a measurement is the difference between the mean and the value of that measurement. The sum of the squared deviations for all the measurements obtained divided by the number of measurements $n$ gives the mean of the squares of the deviations. The

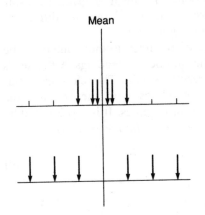

**Fig. 1.3**   Spread of results

square root of this quantity is the root-mean-square deviation or standard deviation $\sigma$.

$$\sigma = \sqrt{\left[\frac{(x_1 - \bar{x})^2 + (x_2 - \bar{x})^2 + \ldots (x_n - \bar{x})^2}{n}\right]}$$

$$\sigma = \sqrt{\frac{\Sigma(x_n - \bar{x})^2}{n}} \qquad [7]$$

Because the deviations are squared before being summed the signs of the deviations automatically all become +. A better measure of the variation about the mean is given by using $(n - 1)$ instead of $n$. However, if $n$ is large the difference between $(n - 1)$ and $n$ is not significant.

The mean value of a set of readings is more likely to be close to the true mean if the number of readings in the set is large and the spread of results small. The closeness thus depends on the number of readings $n$ in the set and the standard deviation $\sigma$ of the set of readings. It is represented by what is termed the *standard error of the mean* $\alpha$, where

$$\alpha = \frac{\sigma}{\sqrt{n}} \sqrt{\left[\frac{\Sigma(x - \bar{x})^2}{n(n - 1)}\right]} \qquad [8]$$

**Table 1.1** Example 2

| Resistance ($\Omega$) $x_n$ | Deviation ($\Omega$) $x - \bar{x}$ |
|---|---|
| 99.1 | +0.1 |
| 99.0 | 0.0 |
| 98.7 | −0.3 |
| 99.4 | +0.4 |
| 98.8 | −0.2 |
| 99.0 | 0.0 |
| $\Sigma x_n = 594.0$ | $\Sigma\lvert x_n - \bar{x}\rvert = 1.0$ |
| $\dfrac{\Sigma x_n}{n} = 99.0$ | $\dfrac{\Sigma\lvert x_n - \bar{x}\rvert}{n} = 0.17$ |

**Example 2**

The following results were obtained from measurements of resistance:

99.1 $\Omega$   99.0 $\Omega$   98.7 $\Omega$   99.4 $\Omega$   98.8 $\Omega$   99.0 $\Omega$

What is (a) the arithmetic mean and (b) the mean deviation from the mean?

*Answer*

Table 1.1 shows the steps involved in obtaining the arithmetic mean, the deviations and the mean deviation. The arithmetic mean is 99.0 $\Omega$ and the mean deviation is 0.17 $\Omega$.

**Example 3**

Repeated measurements of a capacitance using a bridge gave the following results:

8.0 µF   8.1 µF   7.9 µF   7.9 µF   8.3 µF   8.4 µF

What is (a) the arithmetic mean and (b) the standard deviation of the results from that mean?

*Answer*

Table 1.2 shows the steps involved in determining the arithmetic mean, the deviations, the squares of the deviations and hence the standard deviation. The arithmetic mean is 8.1 µF and the standard deviation 0.21 µF.

**Table 1.2** Example 3

| Capacitance ($\mu F$) $x_n$ | Deviation ($\mu F$) $x_n - \bar{x}$ | (Deviation in $\mu F$)$^2$ $(x_n - \bar{x})^2$ |
|---|---|---|
| 8.0 | −0.1 | 0.01 |
| 8.1 | 0.0 | 0.00 |
| 7.9 | −0.2 | 0.04 |
| 7.9 | −0.2 | 0.04 |
| 8.3 | +0.2 | 0.04 |
| 8.4 | +0.3 | 0.09 |

$$\Sigma x_n = 48.6 \qquad\qquad \Sigma(x_n - \bar{x})^2 = 0.22$$

$$\frac{\Sigma x_n}{n} = 8.1 \qquad\qquad \sqrt{\frac{\Sigma(x_n - \bar{x})^2}{n - 1}} = 0.21$$

## Probable error

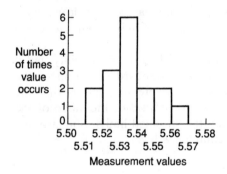

**Fig. 1.4** Histogram

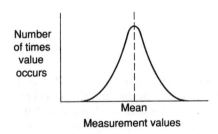

**Fig. 1.5** Gaussian distribution

Consider measurements of some quantity which gave the following results, for 16 measurements:

values between 5.51 and 5.52 come up twice
values between 5.52 and 5.53 come up three times
values between 5.53 and 5.54 come up six times
values between 5.54 and 5.55 come up twice
values between 5.55 and 5.56 come up twice
values between 5.56 and 5.57 come up once

A *histogram* can be plotted showing how the number of times a measurement occurs is related to the values of the individual measurements (Fig. 1.4). The term *frequency* is used for the number of times a measurement occurs and the histogram represents what is called the *frequency distribution*, the area of a bar in the histogram being related to the frequency and the location of the bar to the range of values for which that is the frequency. With a large number of measurement values, when there are random errors the histogram tends to become symmetrical and the midpoints of the histogram bars lie on a smooth, symmetrical, bell-shaped curve (Fig. 1.5). Such a form of distribution is called a *Gaussian* or *normal* distribution.

The frequency distribution of a set of measurements shows how the measured values vary in relation to the mean value and can thus be considered to show the deviations of the various measurements from the mean. Thus for the distribution shown in Fig. 1.5, the most frequent measurement is at the mean, and small deviations from the mean are more likely than large ones. There is also an equal chance of measurements having deviations greater than the mean and ones having deviations less than the mean. As we are dealing with large numbers of measurements, the mean is the true mean and the deviations represent the errors of individual measurements from this true mean.

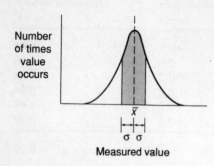

Number of times value occurs

σ σ

Measured value

**Fig. 1.6** Values within 1 σ of the mean

With the histogram in Fig. 1.4 the area of any one bar represents the number of measurements made for the values represented by that bar, the total area of all the bars representing the entire number of values obtained. Similarly the total area under the Gaussian distribution represents the entire number of values obtained. If we draw lines on the distribution to represent all the values lying within one standard deviation of the mean (Fig. 1.6) then the percentage of the total area that lies between those lines is approximately 68%. This means that 68% of all the values obtained lie within one standard deviation of the mean. With the Gaussian distribution, the chance of a measurement occurring within one standard deviation of the mean is 68.3%, within two standard deviations 95.5%, three standard deviations 99.7% and four standard deviations 99.99%. The chance of a measurement falling within 0.6745 σ of the mean is 50%. The 0.6745 σ is called the *probable error*. Thus a statement of the probable error for a set of measurements means that there is a 50% chance that if we take just one measurement, it will have a random error no greater than ±0.6745 σ from the mean value.

If we consider one measurement then the probability that it will lie within plus or minus one standard deviation of the mean is 68.3%. Another way of expressing this is that the limits ±1 σ give a 68.3% *confidence level* of the value of the measurement being between the specified limits. Similarly the limits ±2 σ give a 95.5% confidence level.

**Example 4**

Measurements of the resistance of a batch of resistors gave the following values:

100.1 Ω    101.0 Ω    100.6 Ω    99.5 Ω    99.0 Ω

100.0 Ω    99.5 Ω    100.4 Ω    99.8 Ω    100.1 Ω

Assuming that only random errors are present, what is (*a*) the arithmetic mean, (*b*) the standard deviation and (*c*) the probable error?

*Answer*

Table 1.3 shows the steps involved in determining the arithmetic mean, the deviations and the standard deviation. The arithmetic mean is 100.0 Ω and the standard deviation 0.58 Ω. Thus the probable error is ±0.6745 × 0.58 = ±0.39 Ω. Thus if any one of the resistors is considered, there is a 50% chance that it will have a value within 0.39 Ω of the mean.

**Example 5**

A batch of resistors has an arithmetic mean of 10.0 Ω and a standard

**Table 1.3** Example 4

| Resistance ($\Omega$) | Deviation ($\Omega$) | (Deviation in $\Omega)^2$ |
|---|---|---|
| 100.1 | +0.1 | 0.01 |
| 101.0 | +1.0 | 1.00 |
| 100.6 | +0.6 | 0.36 |
| 99.5 | −0.5 | 0.25 |
| 99.0 | −1.0 | 1.00 |
| 100.0 | 0.0 | 0.00 |
| 99.5 | −0.5 | 0.25 |
| 100.4 | +0.4 | 0.16 |
| 99.8 | −0.2 | 0.04 |
| 100.1 | +0.1 | 0.01 |

$\Sigma x_n = 1000.0$ $\qquad\qquad$ $\Sigma(x_n - \bar{x})^2 = 3.08$

$\dfrac{\Sigma x_n}{n} = 100.0$ $\qquad\qquad$ $\sqrt{\dfrac{\Sigma(x_n - \bar{x})^2}{n - 1}} = 0.58$

deviation of $0.2\,\Omega$. In a batch of 1000, how many might be expected to have resistances between (a) $9.80\,\Omega$ and $10.2\,\Omega$ and (b) $9.60\,\Omega$ and $10.4\,\Omega$?

*Answer*

Assuming that the resistances vary in a completely random manner and thus follow the Gaussian distribution, then the chance of a resistance being within $\pm 1\sigma$ of the mean is 68.3% and within $\pm 2\sigma$ is 95.5%. The higher the probability of an event occurring, the greater the frequency with which it occurs. Thus the chance of a single coin falling heads uppermost is 50%, i.e., 1 in 2. If 1000 coins fall then it is likely that 50% of them will fall heads uppermost. Thus out of a batch of 1000 resistors we can expect 68.3%, i.e., 683, to have resistances within one standard deviation of the mean and 95.5%, i.e., 955, within two standard deviations. Thus for (a), which is one standard deviation from the mean, the answer is 683 and for (b), which is two standard deviations, 955.

**Summation of errors**

A quantity may be determined as a result of calculations carried out on the results of a number of measurements, each of which has some error associated with it. For example, consider the combining of two resistors in series when one is $100 \pm 10\,\Omega$ and the other $50 \pm 5\,\Omega$. The resistances will thus be between 90 and $110\,\Omega$ and between 45 and $55\,\Omega$. The worst possible values of the combined resistors will be when both are at their minimum values and both at their maximum values. Then the combined resistances are $90 + 45\,\Omega$ and $110 + 55\,\Omega$. The combined resistance would thus lie between 135 and $165\,\Omega$. In the absence of error the combined resistance would be $100 + 50 = 150\,\Omega$. Thus the error of the combined resistance is $\pm 15\,\Omega$. This is just the sum of the errors in the two separate resistances, i.e., $\pm 10 \pm 5 = \pm 15\,\Omega$. This represents the worst possible error that could occur, since both resistances would have to be at their minimum or both at their maximum values. A more likely error would be less than this (see equation [10]).

If the quantity $X$ is obtained by adding together the results of the measurements of two quantities $A$ and $B$, then in the absence of any errors in the measurements

$$X = A + B$$

However, if the error in $A$ is $\pm \Delta A$ and that in $B$ is $\Delta B$ then there will be an error in $X$ of $\pm \Delta X$ where

$$X \pm \Delta X = (A \pm \Delta A) + (B \pm \Delta B)$$

Thus, since $X = A + B$,

$$\pm \Delta X = \pm \Delta A \pm \Delta B$$

The error in $X$ will be the greatest when the errors in $A$ and $B$ are the same sign, and so

$$\Delta X = \pm(\Delta A + \Delta B) \qquad\qquad [9]$$

The result of adding the two measurements is to add the errors.

Equation [9] gives what would be the very worst error, because the errors in $A$ and $B$ are unlikely both to be at their maximum values with the same sign at the same time. A more realistic error is given by

$$\Delta X = \pm \sqrt{[(\Delta A)^2 + (\Delta B)^2]} \qquad\qquad [10]$$

When the quantity $X$ is obtained as a result of multiplying two measurements $A$ and $B$, then in the absence of errors

$$X = AB$$

If $\pm \Delta A$ is the error in $A$ and $\pm \Delta B$ the error in $B$ then the error in $X$ of $\pm \Delta X$ is given by

$$X \pm \Delta X = (A \pm \Delta A)(B \pm \Delta B)$$

$$= AB \pm A\Delta B \pm B\Delta A \pm \Delta A\,\Delta B$$

Since $\Delta A$ and $\Delta B$ can be assumed to be small, the product $\Delta A\,\Delta B$ is negligible. Thus

$$X \pm \Delta X = AB \pm A\Delta B \pm B\Delta A$$

$$\Delta X = \pm A\Delta B \pm B\Delta A$$

This can be written as

$$\frac{\Delta X}{X} = \frac{\pm A\Delta B \pm B\Delta A}{AB}$$

$$= \pm \frac{\Delta B}{B} \pm \frac{\Delta A}{A}$$

In the worst possible case both errors have the same sign and so

$$\frac{\Delta X}{X} = \pm \left[ \frac{\Delta B}{B} + \frac{\Delta A}{A} \right]$$

$$\frac{\Delta X}{X} \times 100 = \pm \left[ \frac{\Delta B}{B} \times 100 + \frac{\Delta A}{A} \times 100 \right] \qquad [11]$$

The percentage error in $X$ is the sum of the percentage errors in the measurements. This is the worst possible error when both the error terms have their maximum values and the same sign at the same instant. A more realistic error is

$$\frac{\Delta X}{X} = \pm \sqrt{\left[ \left( \frac{\Delta A}{A} \right)^2 + \left( \frac{\Delta B}{B} \right)^2 \right]} \qquad [12]$$

The above type of derivation can be used to determine the errors when measurements are combined in other ways. When the result is obtained by:

1   Adding measurements: add the errors to obtain the overall worst possible error. A more realistic error is given by

$$\Delta X = \pm \sqrt{[(\Delta A)^2 + (\Delta B)^2]}$$

2   Subtracting measurements: add the errors to obtain the overall worst possible error. A more realistic error is given by

$$\Delta X = \pm \sqrt{[(\Delta A)^2 + (\Delta B)^2]}$$

3   Multiplying measurements: add the percentage errors to obtain the overall percentage error. A more realistic error is

$$\frac{\Delta X}{X} = \pm \sqrt{\left[ \left( \frac{\Delta A}{A} \right)^2 + \left( \frac{\Delta B}{B} \right)^2 \right]}$$

4   Dividing measurements: add the percentage errors to obtain the overall percentage error. A more realistic error is

$$\frac{\Delta X}{X} = \pm \sqrt{\left[ \left( \frac{\Delta A}{A} \right)^2 + \left( \frac{\Delta B}{B} \right)^2 \right]}$$

5   The measurement as a power: multiply the percentage error of the measurement by the power to obtain the overall percentage error. A more realistic error is

$$\frac{\Delta X}{X} = \pm \sqrt{\left[ \left( \frac{\Delta A}{A} \right)^2 + \left( \frac{n \Delta B}{B} \right)^2 \right]}$$

**Example 6**

The power dissipated by a resistor is given by $P = V^2/R$. What will be the worst possible error in the power if the error in the measurement of the voltage $V$ is $\pm 4\%$ and the resistance has a tolerance of $\pm 10\%$?

*Answer*

The percentage error in the $V^2$ term is twice that in $V$ and so is $\pm 8\%$. The percentage error in $P$ is the sum of the errors in $V^2$ and $R$ and so is $\pm 18\%$.

**Best straight line**

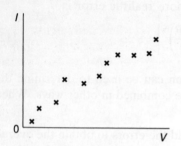

**Fig. 1.7**  Scatter diagram

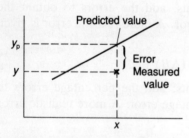

**Fig. 1.8**  Error due to a prediction

Often measurements are made to determine a relationship between two quantities. For example, will a graph of current against potential difference give a straight line graph? However, the measurements will have associated errors and joining up the measurement points on a graph may not give a straight line graph. The resulting sets of plotted points may show some scatter; indeed, a graph showing the positions of the measurement points is often referred to as a *scatter diagram* (Fig. 1.7). The problem is then to decide whether there is a straight line relationship and, if so, what is the best estimate of the straight line.

Suppose we draw a line through the scatter points. For any one point we will have some error between the measured value $y$ and the value predicted $y_p$ by the line (Fig. 1.8). The error is $(y_p - y)$ for the value $x$. The best line is the one for which the sum of the squares of the errors for all the measured points is a minimum, i.e.,

$$\Sigma(y_p - y)^2 = \text{a minimum} \qquad [13]$$

The process of finding the best line is thus effectively the drawing of lines and determining the sums of the squares of the errors for each line until we end up with the line giving the minimum condition. The term *regression line* is used for the line found by using this condition and the method is called *least squares regression* (note that the word regression has no particular significance now, the original reason for the term being no longer relevant). In the above we have assumed that the $x$ values are without error and only the $y$ values of measurements are in error. This gives us what is termed the regression line for predicting $y$ from $x$. We could have assumed that the $y$ values were without error and only the $x$ values are in error. We would then have ended up with the regression line for predicting $x$ from $y$ (Fig. 1.9).

For a series of $n$ measurement points $(x_1, y_1)$, $(x_2, y_2)$, . . ., $(x_n, y_n)$ the gradient $m$ of the best straight line is given by

$$m = \frac{n\Sigma(xy) - \Sigma x \Sigma y}{n\Sigma(x^2) - (\Sigma x)^2} \qquad [14]$$

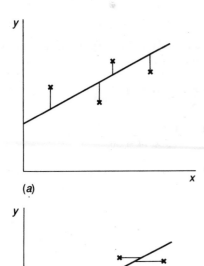

(a)

(b)

**Fig. 1.9** Regression lines:
(a) predicting y from x, (b) predicting x
from y

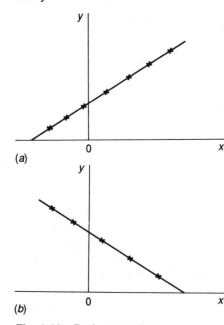

(a)

(b)

**Fig. 1.10** Perfect correlation:
(a) r = +1, (b) r = −1

and the intercept $c$ with the $y$ axis by

$$c = \frac{\Sigma y (\Sigma x^2) - \Sigma x \Sigma (xy)}{n \Sigma (x^2) - (\Sigma x)^2}$$ [15]

or

$$c = \frac{\Sigma y - m \Sigma y}{n}$$ [16]

The best straight line has then the equation

$$y = mx + c$$ [17]

The above analysis leads to the identification of the best straight line that can be drawn through a scatter plot; it does not, however, give any clue as to whether the straight line relationship is a likely one for the set of measurements. How certain can we be that the relationship is given by the straight line? To do this we need to calculate what is termed the *correlation coefficient r*.

$$r = \frac{(1/n) \Sigma [(x - \bar{x})(y - \bar{y})]}{\sigma_x \sigma_y}$$ [18]

where $\bar{x}$ is the mean of the $x$ values and $\bar{y}$ that of the $y$ values, $\sigma_x$ the standard deviation of the $x$ values and $\sigma_y$ that of the $y$ values. The value of $r$ lies between +1 and −1. A value of +1 denotes perfect correlation between $y$ and $x$ with an increase in $y$ resulting in an increase in $x$. Perfect correlation means a straight line graph since one quantity is directly related to the other (Fig. 1.10(a)). A value of −1 denotes perfect correlation between $y$ and $x$ with an increase in $y$ resulting in a decrease in $x$. There is again a straight line relationship between the two quantities concerned (Fig. 1.10(b)). When $r = 0$ there is no correlation between $y$ and $x$. The points are completely randomly scattered and $y$ does not depend on $x$ in any way (Fig. 1.11). The closer $r$ is to +1 or −1 the more the points cluster about a straight line.

With $r = +1$ or −1 the two regression lines obtained by predicting $y$ from $x$ and from predicting $x$ from $y$ coincide. When $r = 0$ the two regression lines are at right angles to each other, each line being parallel to the coordinate axes. With $r$ having values between 0 and +1 or −1 there is some angle between the lines, the closer $r$ is to +1 or −1, the smaller the angle is.

**Example 7**

Find the best straight line for predicting $y$ from $x$ for the following measurement points:

$$x = 1, y = 2; \ x = 2, y = 4; \ x = 3, y = 5$$

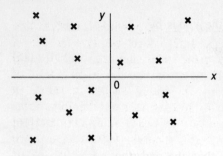

**Fig. 1.11** Complete lack of correlation, $r = 0$

**Table 1.4** Example 7

| $x$ | $y$ | $xy$ | $x^2$ |
|---|---|---|---|
| 1 | 2 | 2 | 1 |
| 2 | 4 | 8 | 4 |
| 3 | 5 | 15 | 9 |

$\Sigma x = 6$  $\Sigma y = 11$  $\Sigma xy = 25$  $\Sigma x^2 = 14$

*Answer*

This can be tackled using equations [14] and [15]. Table 1.4 shows steps in the calculation. Thus for $n = 3$

$$m = \frac{n\Sigma(xy) - \Sigma x \Sigma y}{n\Sigma(x^2) - (\Sigma x)^2} = \frac{3 \times 25 - 6 \times 11}{3 \times 14 - 36} = 1.5$$

$$c = \frac{\Sigma y(\Sigma x^2) - \Sigma x \Sigma(xy)}{n\Sigma(x^2) - (\Sigma x)^2} = \frac{11 \times 14 - 6 \times 25}{3 \times 14 - 36} = 0.67$$

Thus the best line has the equation

$$y = 1.5x + 0.67$$

**Example 8**

Determine the correlation coefficient for the data given in Example 7.

*Answer*

Table 1.5 shows the steps in the calculation. The correlation coefficient is given by equation [18] as

$$r = \frac{(1/n)\Sigma[(x - \bar{x})(y - \bar{y})]}{\sigma_x \sigma_y} = \frac{(1/3) \times 3.0}{1.0 \times 2.34} = 0.43$$

**Table 1.5** Example 8

| $x$ | $y$ | $x - \bar{x}$ | $(x - \bar{x})^2$ | $y - \bar{y}$ | $(y - \bar{y})^2$ | $(x - x)(y - y)$ |
|---|---|---|---|---|---|---|
| 1 | 2 | −1.0 | 1.0 | −1.7 | 2.89 | +1.7 |
| 2 | 4 | 0.0 | 0.0 | −0.3 | 0.09 | 0.0 |
| 3 | 5 | +1.0 | 1.0 | +1.3 | 1.69 | +1.3 |
| $\Sigma = 6$ | $\Sigma = 11$ | | $\Sigma = 2.0$ | | $\Sigma = 4.67$ | $\Sigma = +3.0$ |
| $\bar{x} = 2$ | $\bar{y} = 3.7$ | | $\sigma_x = 1.0$ | | $\Sigma_y = 2.34$ | |

**Making the curved straight**

Straight line graphs between two quantities $y$ and $x$ have an equation of the form

$$y = mx + c$$

where $m$ is the gradient of the graph and $c$ the intercept on the $y$ axis. Thus when measured data give a straight line graph it is relatively easy to fit an equation to the relationship between the quantities measured. This applies even when the measured points have errors. When the relationship does not give a straight line it is much more difficult. The relationship $y = x^2$ gives a non-linear graph (Fig. 1.12(a)). However, if we take logarithms of both sides of the equation then

$$\lg y = 2 \lg x$$

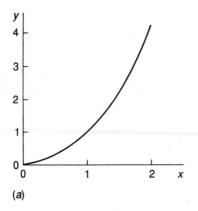

(a)

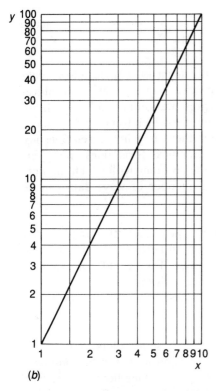

(b)

**Fig. 1.12** $y = x^2$, (a) on linear–linear graph axes, (b) on log–log graph axes

**Fig. 1.13** $y = 2^x$, (a) on linear–linear graph axes, (b) on log–linear graph axes

Thus if $\lg y$ is plotted against $\lg x$, the graph is a straight line with a gradient of 2 (Fig. 1.12(b)). If we have a set of measurements which we think will obey the relationship $y = x^2$, it is easier to see if this is the case if a graph is used of $\lg y$ against $\lg x$, than if $y$ is plotted against $x$. Now in plotting log graphs there is nothing to stop the logarithms being found for each value of $y$ and each value of $x$ and the graph plotted on ordinary graph paper. However, a more convenient way is to use what is called lg–lg graph paper which automatically does the conversion for you.

There are many relationships between quantities of the form $y = a^x$. A graph of such a relationship on linear–linear graph paper is not a straight line (Fig. 1.13(a)). However, if logarithms are taken

$$\lg y = x \lg a$$

A graph of $\lg y$ against $x$ will produce a straight line graph with a gradient of $\lg a$. Again, the logarithms can be obtained and then conventional graph paper used or lg–linear graph paper used which does the job automatically for you (Fig. 1.13(b)). Such a plot is often referred to as a semi-log plot.

Thus by choosing the form of the divisions on the axes of the graph paper, relationships which would otherwise give curves can be forced to give straight lines.

**Example 9**

Measurements of the current through a resistor and the potential difference across it gave the following results:

| Current (mA) | 100 | 200 | 300 | 400 |
|---|---|---|---|---|
| Potential difference (V) | 1.0 | 4.0 | 9.0 | 16.0 |

The law relating the current $I$ and the potential difference $V$ is suspected to be of the form $V = aI^n$. If this is the case, what graph axes should be used to give a straight line graph? Hence plot such a graph and determine $n$.

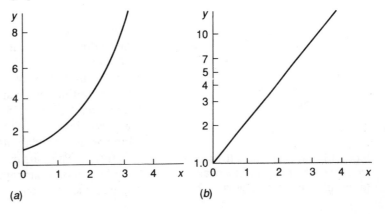

(a)

(b)

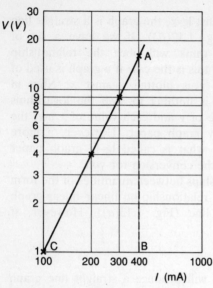

**Fig. 1.14** Example 9

*Answer*

Taking logarithms

$$\lg V = n \lg I + \lg a$$

Thus a graph of $\lg V$ against $\lg I$ will be a straight line graph with a gradient of $n$. Figure 1.14 shows the graph. Since it is a straight line the law is valid. The gradient of the graph is $AB/BC = (\lg 16 - \lg 1)/(\lg 400 - \lg 100) = 2.0$.

## Data display

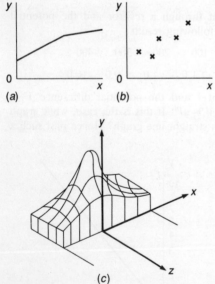

**Fig. 1.15** (a) Line graph, (b) scatter plot, (c) contour plot, (d) bar chart, (e) pie diagram, (f) histogram

The presentation of data in the form of a table allows data values to be recorded in a precise way. However, graphical displays provide a pictorial representation of results which are more readily comprehended and for which trends can be more easily discerned, e.g., whether there is a linear relationship. The presentation of data in graphical form does, however, involve some compromises in the accuracy with which the data can be displayed. There are a number of ways data can be graphically displayed.

1   *Line graphs* (Fig. 1.15(*a*))   These show the trend of one variable quantity with respect to another, there being two continuously variable quantities. They could, for example, be the variation of resistance with time.

2   *Scatter plots* (Fig. 1.15(*b*)   These show the data points on graphical axes when there are two continuosly variable quantities, the scatter of the points thus being evident.

3   *Contour plots* (Fig. 1.15(*c*))   These are plots representing three-dimensions. They can be considered to be composed of a large number of line graphs showing how $y$ varies with $x$ for a multitude of different values of $z$. There are thus three continuously varying quantities.

4   *Bar charts* (Fig. 1.15(*d*))   These compare the values of one continuous variable in a number of situations, e.g., the price of computers produced by different manufacturers. The bars can be drawn vertically or horizontally, the length of a bar representing the size of the variable. In some cases the bars might be drawn in the form of the

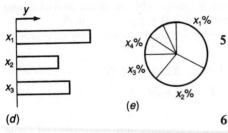

(d)

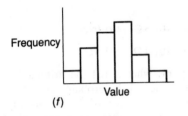

(e)

(f)

**Fig. 1.15** continued

variable, e.g., different height computers being used to represent the price.

5  *Pie charts* (Fig. 1.15(e))  These show the parts that make up an entity when there is just one variable; the area of the entire circular pie representing 100% and the area of each slice representing the percentage contribution of each item.

6  *Histograms* (Fig. 1.15(f))  Histograms are bar charts where the length of a bar represents the frequency with which some value or range of values occurs, i.e., there is just one continuous variable quantity.

**Example 10**

Select a form of graphical display for the following sets of data:

(a)  The percentage of resistors produced by a machine within the 2%, 5%, 10% and greater than 10% tolerance bands.

(b)  The values of the resistances produced by a machine during a 10 minute test period.

(c)  The variation of the mean resistances of a batch of resistors with time.

*Answer*

(a)  Since only the resistance varies between the resistors, and the results are expressed as a percentage, a good form of display is the pie chart.

(b)  This has only one variable, resistance, and needs to portray the distribution of the resistance values. Hence a histogram can be used.

(c)  This has two continuously variable quantities, resistance and time. Hence a line graph or scatter plot can be used.

**Problems**

1  Explain what is meant by random errors and systematic errors and give examples of each.

2  An ammeter is used to measure a current and gives a reading of 120 mA when the true current is 125 mA. What is (a) the error and (b) the percentage error?

3  Distinguish between accuracy and precision.

4  Measurements of a resistance over a period of time gave the following results:

$$10.16\,\Omega \quad 10.15\,\Omega \quad 10.14\,\Omega \quad 10.05\,\Omega \quad 10.00\,\Omega \quad 9.99\,\Omega$$

What is the (a) the arithmetic mean, (b) the mean deviation and (c) the standard deviation for the resistance over that period of time?

5  Measurements of a voltage over a period of time gave the following results:

$$7.0\,V \quad 8.5\,V \quad 9.0\,V \quad 9.0\,V \quad 10.0\,V \quad 10.5\,V$$

What is (a) the arithmetic mean, (b) the mean deviation and (c) the standard deviation over that period of time?

6 The breakdown voltages for 10 lengths of wire were measured as being:

2.5 kV   2.7 kV   3.0 kV   3.2 kV   2.6 kV

2.9 kV   3.1 kV   3.1 kV   2.6 kV   2.8 kV

What is the arithmetic mean and the standard deviation?

7 The resistances of a batch of resistors were measured and found to be

101.7 Ω   101.2 Ω   101.3 Ω   101.5 Ω   101.0 Ω

101.1 Ω   101.3 Ω   101.2 Ω   101.4 Ω   101.3 Ω

If only random errors are present, what is (a) the arithmetic mean, (b) the standard deviation and (c) the probable error?

8 A batch of resistors has a mean resistance of 50.0 Ω and a standard deviation of 1.0 Ω. Assuming that there are only random errors, what are the resistance deviations from the mean at which there will be only 3 in 1000 resistors with greater deviations?

9 The capacitances of a batch of 20 capacitors were measured and found to be:

9.5 μF   9.2 μF   9.6 μF   9.5 μF   9.7 μF   9.4 μF   9.9 μF

9.5 μF   9.6 μF   9.4 μF   9.6 μF   9.7 μF   9.9 μF   10.0 μF

9.6 μF   9.7 μF   9.3 μF   9.8 μF   9.8 μF   9.9 μF

Assuming that only random errors are present, what is (a) the arithmetic mean, (b) the standard deviation and (c) the probable error?

10 The current through a resistor is measured as being $0.10 \pm 0.01$ A and the potential difference across it measured as $5.0 \pm 0.25$ V. What is the power and its worst possible error?

11 Resistors of 100 Ω ±10% and 50 Ω ±5% are connected (a) in series and (b) in parallel. What are the total resistances and worst possible errors?

12 The current through a resistor is measured as $2.00 \pm 0.01$ A. If the resistance is specified as $100 \pm 0.2$ Ω, what is the power and its worst possible error (power = $I^2R$)?

13 The resistance of an unknown resistor is determined using a Wheatstone bridge as being $R_1R_2/R_3$, where $R_1 = 100$ Ω ±1%, $R_2 = 120$ Ω ±1% and $R_3 = 50$ Ω ±0.5%. What is the value of the unknown resistance and its worst possible error?

14 Find the best straight line for predicting $y$ from $x$ for the following measurement points:

$x = 1, y = 5; x = 2, y = 8; x = 3, y = 9; x = 4, y = 10$

15 Find the best straight line for predicting $y$ from $x$ for the following measurement points:

$x = 1, y = 1; x = 2, y = 2; x = 3, y = 6; x = 4, y = 7$

16  What are the correlation coefficients in problems 14 and 15?

17  Find the regression lines and the correlation coefficient for the following measurement points:

$$x = 1, y = 1; x = 2, y = 5; x = 3, y = 6;$$

$$x = 4, y = 10; x = 5, y = 14; x = 6, y = 18$$

18  Find the regression lines and the correlation coefficient for the following measurement points:

$$x = 1, y = 1; x = 2, y = 1; x = 3, y = 6,$$

$$x = 4, y = 7; x = 5, y = 10; x = 6, y = 16$$

19  Which form of graph paper linear – linear, log–log or log–linear – should be used if the following relationships are to give straight line graphs?

(a)  $y = 2x^3$,
(b)  $y = 2x$,
(c)  $y = 2^{-3x}$

20  The resistance $R_t$ of a thermistor at temperature $t$ is related to the temperature by the relationship $R_t = ae^{\beta/t}$. What is the form of this relationship which will give a straight line relationship and what form of graph paper should thus be used to give a linear graph?

21  The periodic time $T$ for the oscillation of a pendulum is measured for different pendulum lengths $l$ and the following results obtained:

| $l$ (m) | 0.25 | 0.50 | 0.75 | 1.00 | 1.25 |
|---------|------|------|------|------|------|
| $T$ (s) | 1.00 | 1.42 | 1.74 | 2.01 | 2.24 |

The relationship between the length and the periodic time is believed to be of the form $l = kT^n$. Show that the law is valid by plotting a suitable form of graph to give a straight line. Hence determine the values of $k$ and $n$.

22  The resistance of uniform cross-section wire is given by $R = \varrho L/A$ where $\varrho$ is the resistivity and constant for a given material, $L$ is the length of the wire and $A$ the cross-sectional area. Since $A = \frac{1}{4}\pi d^2$, where $d$ is the diameter of the wire, what should be the axes of graphs plotted to give straight line graphs for (a) the resistance in terms of the diameter and (b) the resistance in terms of the length?

23  Select a form of display for the following:

(a)  Measurements showing how the insulation resistance varies with time.
(b)  The cost breakdown for ATE as 50% capital cost, 30% cost of programming, 10% operation and maintenance cost and 10% operator training cost.
(c)  The spread of resistance values within a batch of resistors.
(d)  The relative costs of meters from different manufacturers.
(e)  The numbers of hours of continuous operation without failure of a particular component.
(f)  The number of customers for a particular instrument in each area of the country.

# 2 Performance specifications

## Introduction

The term *instrument* is used for a device that determines the value or size of some quantity. Thus it could be a ruler used for the determination of the length of some object, a moving coil galvanometer used for determining the size of a current, or perhaps a cathode ray oscilloscope used to determine the wave form of some signal. The act of determining such quantities is called *measurement*. However, whatever the form of instrument, in order to know whether it is capable of making a particular measurement it is necessary to know the instrument specification. The *specification* is a list which gives information detailing what performance can be expected of the instrument and the conditions under which the performance is guaranteed. This book is a consideration of electric and electronic measurements and this chapter is about the terms used in the specifications of instruments used for such measurements.

Specifications are also used to specify the procedures that have to be followed to test whether an instrument, device or process is performing to its specification, and this is considered in Chapter 17.

## Specification terms

There are many terms used in specifications, the following being some of the main ones. Others are introduced at later points in the book and a glossary of such terms is given in Appendix B at the end of the book. The terms can be considered to fall into three main categories: those concerned with the design of the instrument, those with its sensitivity and accuracy and those with its calibration.

The *design terms* are as follows:

1 *Indicating instrument* This is a measuring instrument in which the value of the measured quantity is visually indicated, e.g., a pointer position on a scale. A moving coil ammeter is an example of such an instrument.

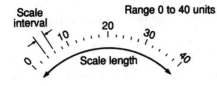

**Fig. 2.1** A scale

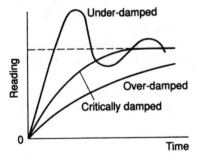

**Fig. 2.2** Dead space

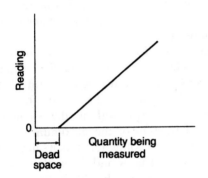

**Fig. 2.3** Damping

2 *Recording instrument* This is a measuring instrument in which the values of the measured quantity are recorded on a chart, e.g., a chart recorder (see Chapter 8).

3 *Scale* The scale on an indicating instrument is the array of marks, together with associated figuring, against which the position of a pointer, light spot, liquid surface or other form of index is indicated (Fig. 2.1).

4 *Scale length* This is the distance, measured along a line which defines the path of the index or pointer, between the end marks of the scale (Fig. 2.1).

5 *Scale interval* The scale interval is the amount by which the measured quantity changes in the index moving between adjacent scale marks (Fig. 2.1).

6 *Range* The range or span of an instrument is the limits between which readings can be made (Fig. 2.1).

7 *Dead space* The dead space of an instrument is the range of values of the quantity being measured for which it gives no reading (Fig. 2.2). Where it occurs it is often at the beginning of the range. Thus an instrument might require a certain value of current before it begins to operate and indicate any current as being present.

8 *Damping* When there is an input to an instrument the index will not generally move straight to the appropriate value on the scale but may overshoot it and then oscillate about the value before finally settling on it (Fig. 2.3). If there was no damping, the oscillations would continue indefinitely. An instrument is said to be damped when the amplitude of the oscillations of the index are progressively reduced or completely suppressed. The term under-damping is used when the degree of damping is such that oscillations of the index occur, the term critical damping when the amount of damping is just sufficient to prevent oscillation, and the term over-damping when the degree of damping is more than sufficient to prevent oscillation. An instrument is said to be *aperiodic* when the motion of its index is critically damped or over-damped.

9 *Sample rate* Some instruments, e.g., digital voltmeters, take samples of the quantity being measured at regular intervals. The greater the sample rate, i.e., the greater the number of samples taken per second, the more readily the instrument readings mirror a rapidly changing input.

The terms for *sensitivity and accuracy* are:

1 *Sensitivity* The sensitivity of an instrument is the relation between the change in reading indicated by movement of the index and the change in the measured quantity that produces it. It can be expressed as:

$$\text{sensitivity} = \frac{\text{change in instrument scale reading}}{\text{change in the quantity being measured}}$$

For example, the sensitivity of an oscilloscope to a $Y$-input may be expressed as 0.05 div/V. Sometimes, however, the reciprocal of this quantity, i.e., 20 V/div. is quoted. In the case of some measurement systems, e.g., a bridge, this definition is expressed in terms of the change in output resulting from a change in the input,

$$\text{sensitivity} = \frac{\text{change in the output}}{\text{change in the input}}$$

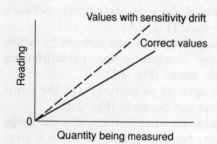

**Fig. 2.4** Sensitivity drift

2 *Sensitivity drift* The sensitivity drift is the amount by which the sensitivity changes as a result of changes in environmental conditions (Fig. 2.4).

3 *Resolution* The resolution or *discrimination* of an instrument is the smallest change in the quantity being measured, i.e., the input to the measurement system, that will produce an observable change in the reading of the instrument. It is sometimes expressed as the smallest change in the input as a fraction of the maximum input value. Thus, for example, a voltmeter might have a fractional resolution of $10^{-4}$. This would mean that on a range of 200 V the resolution would be $200 \times 10^{-4}$ V or 20 mV.

4 *Drift* An instrument is said to show drift if there is a gradual change in its reading over a period of time which is unrelated to any change in input.

5 *Zero drift* This term is used to describe the change in the zero reading of an instrument that can occur with time (Fig. 2.5).

6 *Accuracy* The accuracy of an instrument is the extent to which the reading it gives might be wrong, i.e., the extent to which it differs from the true value. The true value is the actual value assigned to a quantity, this being the value as indicated by the best standard measuring instrument. Accuracy can be quoted as plus or minus some value of the quantity being measured, e.g., an ammeter might be quoted as having an accuracy of ±0.1 A at some particular current value or for all its readings. This means that the value indicated by the meter is only guaranteed to be within + or −0.1 A of the true value. Alternatively, accuracy can be quoted as a percentage of the measured quantity, e.g., ±4%. This would indicate that the value indicated by the instrument is only guaranteed to be within + or −4% of the true value. Thus if the instrument gave a reading of 5.0 A then the true value is within + or −4% of 5.0 A or + or −0.2 A. Alternatively, the accuracy can be quoted as a percentage of the full-scale deflection (f.s.d.)

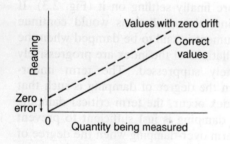

**Fig. 2.5** Effect of zero drift

of the instrument, e.g., an ammeter might have an accuracy quoted as ±2% f.s.d. This means that the accuracy of a reading of the ammeter when used for any reading within the range 0–10 A is plus or minus 2% of 10 A, i.e., + or −0.2 A. The term *fiducial value* is used for the quantity to which reference is made in order to define the accuracy as a fraction or percentage. Generally the fiducial value is the full-scale deflection.

In the case of digital instruments the accuracy is generally quoted as ± the percentage of the reading ±1 digit. Thus, for example, if the accuracy of a digital voltmeter was quoted as ±0.5% of reading ±1 digit then if the reading was 100 V and 1 digit was 1 V the accuracy is ±0.5% of 100 V ±1 V, i.e., ±6 V.

7 *Error* The error is the difference between the result of the measurement and the true value of the quantity being measured:

$$\text{error} = \text{measured value} - \text{true value}$$

Thus if the measured value is 2.2 A when the true value is 2.0 A the error is +0.2 A. If the measured value had been 1.8 A then the error would have been −0.2 A. Errors are sometimes quoted as the fractional error, i.e., the error as a fraction of the true value:

$$\text{fractional error} = \frac{\text{measured value} - \text{true value}}{\text{true value}}$$

or as a percentage fractional error:

$$\text{percentage error} = \frac{\text{measured value} - \text{true value}}{\text{true value}} \times 100\%$$

8 *Zero error* The term zero error is used for the value indicated by the index when there is zero input to the instrument (Fig. 2.5). For electronic systems the term *offset* is used, the offset being the deviation of the output signal from zero when the input is zero.

9 *Bias* The bias of an instrument is the constant error that exists for the full range of its measurements.

10 *Non-linearity error* A linear relationship for an element or a system means the output is directly proportional to the input. Thus doubling the input doubles the output, trebling the input trebles the output. For an instrument where there is a linear relationship, the index moves twice as far in going from 0 to 2 units as from 0 to 1 and thus the scale markings for 1, 2, 3, etc. are equally spaced. In many instances, however, though a linear relationship is used, it is not perfectly linear and so errors occur. The non-

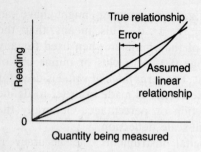

**Fig. 2.6** Non-linearity error

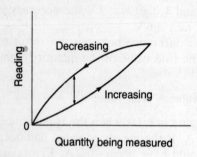

**Fig. 2.7** Hysteresis error

linearity error is the difference between the true value and what is indicated when a linear relationship is assumed (Fig. 2.6). Non-linearity is often expressed in terms of the maximum non-linearity error as a percentage of the full-scale deflection:

$$\text{non-linearity} = \frac{\text{max. non-linearity error}}{\text{full-scale deflection}} \times 100\%$$

11 *Hysteresis error* Instruments can give different readings, and hence an error, for the same value of measured quantity according to whether that value has been reached by a continuously increasing change or a continuously decreasing change (Fig. 2.7). This is called hysteresis and occurs as a result of such things as bearing friction and slack motion in gears in instruments. The hysteresis error is the difference between the measured values obtained when the measured quantity is increasing and when decreasing to that value. Hysteresis is often expressed in terms of the maximum hysteresis as a percentage of the full-scale deflection:

$$\text{hysteresis} = \frac{\text{maximum hysteresis error}}{\text{full-scale deflection}} \times 100\%$$

12 *Reliability* The reliability of an instrument is the probability that it will operate to an agreed level of performance under the conditions specified for its use.

13 *Repeatability* The repeatability of an instrument is its ability to display the same reading for repeated applications of the same value of the quantity being measured.

14 *Reproducibility* The reproducibility of an instrument is its ability to display the same reading when it is used to measure a constant quantity over a period of time or when that quantity is measured on a number of occasions.

15 *Stability* The stability of an instrument is its ability to display the same reading when it is used to measure a constant quantity over a period of time or when that quantity is measured on a number of occasions.

Finally, the terms used for *calibration* are:

1 *Calibration* This is the process of determining the relationship between the value of the quantity being measured and the corresponding position of the instrument index.

2 *Scale factor* For instruments having an arbitrary scale, or multi-range instruments, this is the factor by which the indicator readings have to be multiplied to obtain the value for the measured quantity.

3 *Standard* A standard, whether it be a material standard

or a specification of a test method, provides the method of comparison by which an instrument is calibrated (see the section on standard specification later in this chapter for further discussion).

4  *Traceability*  All standards should be calibrated against superior standards that are traceable to national or international standards (see the section on standard specifications). Generally, calibration of an instrument requires it to be calibrated against a traceable standard which is capable of a 10 to 1 improvement in accuracy. Thus a meter which is required to be capable of providing measurements with an accuracy of ±1% of the full-scale deflection requires calibrating against a standard which has an accuracy better than ±0.1% in the same range.

In addition to the above general specification terms there are other more specific terms such as:

1  *Dimensions and weight*  These are usually the height, width, depth and weight of an instrument.
2  *Operating temperature*  This is the temperature range within which the instrument can be used and give the stated accuracy, stability, etc. quoted in the rest of its specification.
3  *Output impedance*  This is the impedance between the output terminals of the instrument. This has particular significance in relation to loading effects (see Chapter 3 for more details).
4  *Power input*  This is the electrical power input required for the operation of the instrument.
5  *Signal-to-noise ratio*  The signal-to-noise ratio is the ratio of the signal level $V_s$ to the internally generated noise level $V_n$. It is usually expressed in decibels, i.e.,

$$\text{signal-to-noise ratio in decibels} = 20\lg(V_s/V_n)$$

See the next section for a discussion of the decibel unit and Chapter 3 for further discussion of noise and other related terms.

6  *Response time*  A certain time, called the response time, has to elapse when a quantity being measured changes before the measuring instrument responds fully to the change.
7  *Bandwidth*  The readings given for a constant input to an instrument will generally depend on the frequency of the input. Thus, for example, with instruments having pointers, the pointer may just be unable to keep up with the changing signal when the frequency is increased. The term bandwidth is used to give an indication of the range of frequencies over which the instrument can be used. The

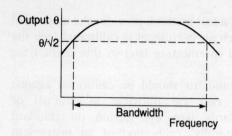

**Fig. 2.8** Bandwidth

bandwidth can be defined as the range of frequencies for which the reading given by the instrument is no less than 70.7% of its peak value θ (Fig. 2.8). The 70.7% of θ is θ/√2. An alternative way of expressing this is that the bandwidth is the range of frequencies for which the reading is within 3 dB of its peak value. A change of 3 dB means a reading which changes by a factor of 1/√2.

$$\text{change in decibels} = 20\lg\left(\frac{\text{value}}{\text{max. value}}\right)$$

$$= 20\lg\left(\frac{1}{\sqrt{3}}\right) = -3$$

**8** *Slew rate* The slew rate is the maximum rate of change with time that the output can have. This term is used, for example, with chart recorders to indicate the maximum rate at which the recorder pen can move across the chart. If the input signal is changing more rapidly then the output will not keep up with it.

**Example 1**

A multimeter as part of its specification includes the following statements:

Ranges
d.c. volts: 100 mV, 3 V, 10 V, 30 V, 100 V, 300 V, 600 V, 1000 V, 3000 V
a.c. volts: 3 V, 10 V, 100 V, 300 V, 600 V, 1000 V, 3000 V
d.c. current: 50 μA, 300 μA, 1 mA, 10 mA, 100 mA, 1 A, 10 A
a.c. current: 10 mA, 100 mA, 1 A, 10 A
resistance: 0–2 kΩ, 0–200 kΩ, 0–20 MΩ

Accuracy at 20 °C
d.c.: ±1% of full-scale deflection
a.c.: ±2% of full-scale deflection at 50 Hz
resistance: ±3% of reading at centre scale

What will be the accuracy of the reading obtained when the meter is used to measure (*a*) a d.c. voltage of 5 V on the 10 V range, (*b*) a d.c. voltage of 5 V on the 30 V range, (*c*) an a.c. current of 5 mA on the 10 mA range, (*d*) a resistance of 50 kΩ on the 0–200 kΩ range? Note that the resistance scale is non-linear and has scale markings, for all ranges, of 0–2000 with a centre scale marking of 20.

*Answer*

(*a*) The accuracy on the 5 V d.c. range is ±1% of the full-scale deflection, i.e., ±1% of 5 V or ±0.05 V. Thus the reading is 5 ± 0.05 V.
(*b*) The accuracy on the 30 V d.c. range is ±1% of the full-scale deflection, i.e., ±1% of 30 V or ±0.3 V. Thus the reading is 5 ± 0.3 V.

(c) The accuracy on the 10 mA a.c. range is ±2% of the full-scale deflection, i.e. ±2% of 10 mA or ±0.2 mA. Thus the reading is 5 ± 0.2 mA.

(d) The accuracy on the 0–200 kΩ range is ±3% of the reading at centre scale. The scale markings would indicate that if the end of scale marking of 2000 is to correspond to 200 kΩ, then the centre scale marking of 20 corresponds to 2 kΩ. Thus the accuracy is ±3% of 2 kΩ or 0.06 kΩ. Thus the reading is 50 ± 0.06 kΩ.

## Example 2

A thermocouple is used with a potentiometer to give a temperature measurement system. The instrument has been calibrated assuming a linear relationship between the e.m.f. produced by the thermocouple and the temperature. The thermocouple gives at its calibration points no e.m.f. at 0°C and 0.645 mV at 100°C, and when used to measure a temperature of 60°C gives 0.365 mV. What is the non-linearity error?

*Answer*

For a linear relationship the e.m.f. produced per degree will be constant over the entire range. This is

$$\text{e.m.f./°C} = \frac{0.645 - 0}{100} = 0.00645 \, \text{mV}$$

Hence an e.m.f. of 0.365 mV should indicate a temperature of

$$\text{temperature} = \frac{0.365}{0.00645} = 56.6 \, °C$$

Since the true temperature is 60°C this is a non-linearity error of 60 − 56.6 = 3.4°C.

## Example 3

An operational amplifier is specified as having: offset voltage 5 mV, slew rate 13 V/μs. Explain the significance of the data.

*Answer*

The offset voltage is the output, in this case 5 mV, that can occur when there is zero input to the amplifier. The slew rate is the maximum rate at which the output can change. Thus if the input signal was changing at a faster rate than 13 V/μs the output would fail to keep up with it. Thus if there was an abrupt change in input the best the amplifier output could do is change at the rate of 13 V/μs, and so would lag behind the input.

## Example 4

A recorder is specified as having a sensitivity of 0.010 mm/μA. What will be the movement of the recorder pen for an input of 1 mA?

*Answer*

Since the sensitivity is

$$\text{sensitivity} = \frac{\text{change in the output}}{\text{change in the input}}$$

Then the change in output is $0.010 \times 1000 = 10\,\text{mm}$.

**Decibels**

Note that the ratio between two values of electric power is usually expressed on a logarithmic scale. With base-10 logarithms the ratio is given the unit the *bel*. Thus power $P_1$ is said to be $N$ bels more than power $P_2$ when

$$N_{\text{bel}} = \lg(P_1/P_2)$$

The *decibel* is one-tenth of a bel and so

$$N_{\text{dB}} = 10\lg(P_1/P_2)$$

Power levels are sometimes compared in decibels to a standard power level of $1\,\text{mW}$. When this happens the power is quoted as being in units of dBm. Thus, for example, a power of $100\,\text{mW}$ can be expressed as

$$10\lg(100/1) = 10 \times 2 = 20\,\text{dBm}$$

This means that the power is $20\,\text{dB}$ above a power of $1\,\text{mW}$. If the power had been less than $1\,\text{mW}$ then the power would have been a negative quantity, e.g., $0.01\,\text{mW}$ is $-20\,\text{dBm}$.

**Example 5**

Express $6\,\text{dB}$ as a power ratio.

*Answer*

For this situation the power ratio $N$ is

$$10\lg N = 6$$

Thus $\lg N = 0.6$ and so

$$N = 10^{0.6} = 3.98$$

Thus, to the nearest integer, $6\,\text{dB}$ means that one power is about four times the other power.

**Example 6**

What power is represented by $-13\,\text{dBm}$?

*Answer*

This means that there is a power of $-13\,\text{dB}$ with reference to a power of $1\,\text{mW}$. The power ratio $N$ is

$$10\lg N = -13$$
$$\lg N = -1.3$$
$$N = 10^{-1.3} = 0.05$$

Thus the power is $0.05 \times 1\,\text{mW} = 0.05\,\text{mW}$.

**Standard specifications**

International and national organizations and high users of items, such as the military, issue standard specifications for the guidance of manufacturers and users of components, equipment and processes. Internationally, general standards are issued by the International Organization for Standards (ISO) while electrotechnical standards are issued by the International Electrotechnical Commission (IEC). At European level there is the European Committee for Standardization (CEN). At national level in Britain there is the British Standards Institute (BSI), in France the Association Française de Normalisation (AFNOR, using for standards the main prefixes NF or UTE), in Germany the Deutsches Institut für Normung (DIN) and in the USA there is the American National Standards Institute (ANSI), the Electronic Industries Association (EIA), the Institute of Electrical and Electronics Engineers (IEEE), the Institute for Interconnecting and Packaging Electronic Circuits (IPC), the Instrument Society of America (ISA) and the National Electrical Manufacturers Association (NEMA). For the military there are standards issued by the British Ministry of Defence (DEF) and American military specifications (MIL and MIL-STD).

The specifications from such organizations give, for example, standard definitions of quantities, symbols for engineering drawings, materials, test procedures, and criteria for selecting, using, calibrating, controlling and maintaining equipment. For example, there are:

1   *British Standards*:
    (*a*) BS 2643 Glossary of terms relating to the performance of instruments.
    (*b*) BS 4739 Method for the expression of the properties of cathode ray oscilloscopes.
    (*c*) BS 5704 Methods for specifying the performance of digital d.c. voltmeters and d.c. electronic analogue-to-digital converters.

2   *International Electrotechnical Commission*:
    (*a*) IEC 50 part 301 General terms on measurements in electricity.
    (*b*) IEC 50 part 302 Electrical measuring instruments.
    (*c*) IEC 50 part 303 Electronic measuring instruments.

3   *Institute of Electrical and Electronics Engineers*:
    (*a*) IEEE 100 Dictionary of electrical and electronic terms.
    (*b*) IEEE 855 Specification for microprocessor operating interfaces.

4   *Deutsches Institut für Normung*:
    (*a*) DIN 2080 Electrical measurement.

(*b*) DIN 2090 Electrical measuring apparatus and instruments.

Instrument manufacturers in their specifications for instruments may thus include such phrases as: the instrument is supplied to the accuracy specified, by BSXX:1970 (the year is often quoted with the standard's reference number to indicate which version of the standard is being applied). This brief statement then indicates without any further detail the accuracy and the conditions under which it can be achieved.

To illustrate this, BS89:1970 groups instruments into classes according to their accuracy, the accuracy being stated as a percentage of the fiducial value. The *fiducial value* is the quantity to which reference is made in order to specify the accuracy of an instrument, it usually being the full-scale value.

| Class | | | | | | | | |
|---|---|---|---|---|---|---|---|---|
| 0.05 | 0.1 | 0.2 | 0.3 | 0.5 | 1.0 | 1.5 | 2.5 | 5.0 |
| Accuracy (%) | | | | | | | | |
| ±0.05 | ±0.1 | ±0.2 | ±0.3 | ±0.5 | ±1.0 | ±1.5 | ±2.5 | ±5.0 |

Thus a manufacturer stating that an instrument is to BS89:1970 class 0.5 means that it has an accuracy of ±0.5% of the fiducial value, under the conditions specified in the standard.

**Example 7**

BS89:1970 specifies that the lower limit of effective range of a permanent magnet moving coil meter is 1/10th of the fiducial value. The effective range is defined as that part of the scale over which measurements can be made to the specified accuracy. Thus if an instrument manufacturer in the specification states that the instrument, range 10 A, is to BS89:1970 class 1.0, what can be stated about the accuracy of the instrument?

*Answer*

Using the data given earlier for the accuracy of different class instruments to BS89:1970, then the specified accuracy is ±1.0% of the fiducial value. Taking this to be the full-scale deflection, then the accuracy is ±0.1 A. This accuracy only, however, applies above 1/10th of 10 A, i.e., above the 1 A mark.

**Quality standards**

Guidance on quality management and three alternative models which can provide a structure for a contractual quality agreement between supplier and purchases are specified by the International Organization for Standards in ISO 9000–9004, and identical specifications by the European Committee for Standardization as EN 29000–29004 and British Standards Institution as BS 5750 (Parts 0–4). A major requirement is that suppliers should establish and maintain effective, economical and demonstrable systems to ensure that materials or services conform to specified requirements. The procedures that have to be followed when selecting, using, calibrating, controlling and maintaining measurement standards and measuring equipment include the requirements for:

1  The supplier to establish and maintain an effective system for the control and calibration of measurement standards and measuring equipment.
2  All personnel performing calibration functions to have adequate training.
3  The calibration service to be periodically and systematically reviewed to ensure its continued effectiveness.
4  All measurements, whether for the purposes of calibration or product assessment, to take into account all the errors and uncertainties in the measurement process.
5  Calibration procedures to be documented.
6  Objective evidence that the measurement system is effective to be readily available to customers.
7  Calibration to be performed to equipment traceable to national standards.
8  A separate calibration record to be kept for each measuring instrument, these demonstrating that all measuring instruments used are capable of performing measurements within the designated limits and including as a minimum:

(a) A description of the instrument and a unique identifier.
(b) The calibration data.
(c) The calibration results.
(d) The calibration interval plus date when the next calibration is due.

and additionally, depending on the type of instrument concerned:

(e) The calibration procedure.
(f) The permissible error limits.
(g) A statement of the cumulative effects of uncertainties in calibration data.
(h) The environmental conditions required for calibration.
(i) The source of calibration used to establish traceability.
(j) Details of any repairs or modifications which might affect the calibration status.
(k) Any use limitations of the instrument.

9  All equipment to be labelled to show its calibration status and any usage limitations.
10  Any instrument which has failed or is suspected or known

to be out of calibration to be withdrawn from use and labelled conspicuously to prevent accidental use.

11 Adjustable devices to be sealed to prevent tampering.

**Problems**

1 An oscillator has as part of its specification the following information:

> Frequency ranges: 10 to 100 Hz, 100 Hz to 1 kHz, 1 to 10 kHz, 10 to 100 kHz, 100 kHz to 1 MHz

> Accuracy: ±3% of full-scale range

What will be the accuracy of a frequency setting of (a) 500 Hz on the 100 Hz to 1 kHz range, (b) 500 kHz on the 100 kHz to 1 MHz ranges?

2 An electronic timer is specified as having an accuracy on the 0–10 ms range of ±2%. What will be the accuracy for a reading of 5 ms on that range?

3 A digital voltmeter, three digits 000 to 999, has an accuracy specification of ±0.2% of the reading ±1 digit. What is the accuracy for (a) a full-scale reading, (b) a one-third scale reading?

4 A resistance thermometer has a resistance $R$ given by

$$R = 100(1 + 4.9 \times 10^{-3}T - 5.9 \times 10^{-7}T^2)$$

where $T$ is the temperature in °C. What will be the non-linearity error at 50 °C if a linear relationship is assumed between 0 °C and 100 °C, i.e., the $T^2$ term is zero?

5 An amplifier is specified as having a temperature sensitivity of 100 μV/°C. What will be the output of the amplifier if the ambient temperature changes by 2 °C?

6 A multimeter includes the following statement in its specification: response time 1 s to full scale. What is the significance of the statement?

7 A chart recorder is said to have a dead-band of ±0.3% of span. What does this mean?

8 A chart recorder is said to have 0.05% repeatability. What does this mean?

9 A digital voltmeter is specified as having a resolution of 1 μV on its 100 mV range. What does this mean?

10 What is the power gain in decibels when the input power to a system is 5 mW and the output power is 25 mW?

11 What is the power gain ratio corresponding to a decibel power gain of (a) +10 dB, (b) −5 dB?

12 The British Standards specification for voltage transformers BS 3941:1975 states that the voltage error shall not exceed values given in a table at any voltage under specified conditions of use such as any voltage between 80% and 120% of the rated voltage. The table includes the following information:

| Accuracy class | 0.1 | 0.2 | 0.5 | 1.0 | 3.0 |
|---|---|---|---|---|---|
| Percentage voltage error | ±0.1 | ±0.2 | ±0.5 | ±1.0 | ±3.0 |

$$\% \text{ loading error} = \frac{I_a - I}{I} \times 100\% = \frac{-R_a}{R + R_a} \times 100\% \quad [1]$$

Thus if an ammeter with resistance $50\,\Omega$ is connected into a circuit having a total resistance of $200\,\Omega$ then the percentage loading error will be $-(50/250) \times 100 = -20\%$.

Connecting a voltmeter across a resistor in order to measure the potential difference has the effect of connecting of the voltmeter resistance in parallel with that of the resistor and so alters the total resistance and hence the potential difference. The act of attempting to make the measurement has modified the potential difference being measured.

Any active network having two terminals A and B to which an electrical load may be connected can be considered to behave as if the network contained a single source of e.m.f. $E_{Th}$ in series with a single impedance $Z_{Th}$ (Fig. 3.2). This is what is known as *Thévenin's theorem* ($E_{Th}$ is the potential difference measured between A and B for the network with no load and $Z_{Th}$ the impedance of the network between A and B when all the sources of e.m.f. within the network have been replaced by their internal impedances). Connecting a load $Z_L$ across the output terminals of an active network is thus equivalent to connecting $Z_L$ across the equivalent Thévenin circuit, as in Fig. 3.2. The current $I$ through $Z_L$ is thus

$$I = \frac{E_{Th}}{Z_{Th} + Z_L}$$

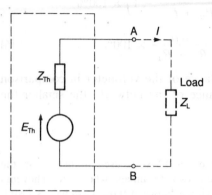

**Fig. 3.2**  Thévenin equivalent circuit

The potential difference across the load $V_L$ is $IZ_L$ and so is

$$V_L = IZ_L = E_{Th}\left(\frac{Z_L}{Z_{Th} + Z_L}\right) \quad [2]$$

The effect of connecting the load across the network is thus to change the potential difference between terminals A and B from $E_{Th}$ to $V_L$. The value of $V_L$ will approach that of $E_{Th}$ the more $Z_L$ is made greater than $Z_{Th}$.

Thus when a voltmeter of resistance $R_m$ is connected across a circuit with a Thévenin equivalent resistance of $R_{Th}$ then the reading indicated by the instrument $V_m$ is

$$V_m = E_{Th}\left(\frac{R_m}{R_m + R_{Th}}\right) \quad [3]$$

where $E_{Th}$ is the Thévenin equivalent voltage of the circuit, i.e., the voltage before the meter was connected. Thus the effect of connecting the voltmeter across the network is to produce an error of $(V_m - E_{Th})$ and thus using equation [2]

$$\text{error} = E_{Th}\left(\frac{R_m}{R_m + R_{Th}}\right) - E_{Th}$$

$$= E_{\text{Th}}\left(\frac{R_m}{R_m + R_{\text{Th}}} - 1\right) \quad [4]$$

The percentage error is

$$\text{percentage error} = \frac{\text{error}}{E_{\text{Th}}} \times 100\%$$

$$= \left(\frac{R_m}{R_m + R_{\text{Th}}} - 1\right) \times 100\%$$

$$= -\frac{R_{\text{Th}}}{R_m + R_{\text{Th}}} \times 100\% \quad [5]$$

Thus the larger the resistance of the voltmeter in comparison with the Thévenin resistance of the network, the smaller the error.

### Example 1

A voltmeter with a resistance of $1\,\text{M}\Omega$ is used to determine the potential difference between two terminals when the Thévenin equivalent resistance of the circuit between those terminals is $2\,\text{M}\Omega$. What is the percentage error resulting from the loading?

*Answer*

Using equation [5]

$$\text{percentage error} = -\frac{R_{\text{Th}}}{R_m + R_{\text{Th}}} \times 100\%$$

$$= -\frac{2}{1 + 2} \times 100\% = -66.7\%$$

### Loading of a potentiometer

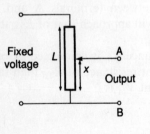

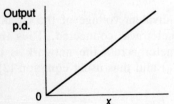

**Fig. 3.3** Potentiometer on open circuit

A potentiometer consists of a resistance track along which a sliding contact can be moved (Fig. 3.3). A fixed potential difference applied between the ends of the resistance track results in a variable potential difference between its sliding contact and one end of the resistance track, the size of the output being then related to the position of the slider along the resistance track. In the figure the potentiometer slider is a distance $x$ from one end of the potentiometer track, the total track length being $L$. If the track has a uniform resistance per unit length, then the open circuit voltage between terminals A and B is $(x/L)V_s$. The open circuit voltage is thus proportional to $x$ and the potentiometer can be said to be linear.

However, when the potentiometer is used with a load (Fig. 3.4) this relationship between the output voltage and $x$ is changed. If we apply Thévenin's theorem, the open circuit voltage is the Thévenin voltage, i.e.,

$$E_{\text{Th}} = (x/L)V_s$$

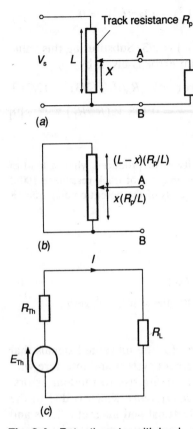

**Fig. 3.4** Potentiometer with load; (a) the circuit, (b) calculation of $R_{Th}$, (c) Thévenin equivalent circuit

The Thévenin impedance is found by replacing $V_s$ with a short-circuit and calculating the impedance between terminals A and B (Fig. 3.4(b)). This consists of two resistors in parallel, one due to the length $x$ and one the length $(L - x)$. If the track has a resistance of $R_p$ then the resistance per unit length is $R_p/L$ and so the two resistances in parallel are $(L - x)(R_p/L)$ and $x(R_p/L)$. Thus the Thévenin resistance $R_{Th}$ is given by

$$\frac{1}{R_{Th}} = \frac{1}{(x/L)R_p} + \frac{1}{[(L - x)/L]R_p}$$

Hence

$$R_{Th} = R_p(x/L)[1 - (x/L)]$$

Figure 3.4(c) shows the Thévenin equivalent circuit for the loaded potentiometer.

The current $I$ in the circuit is given by

$$E_{Th} = I(R_{Th} + R_L)$$

Hence

$$(x/L)V_s = I\{R_p(x/L)[1 - (x/L)] + R_L\}$$

The potential difference across the load is $IR_L$ and is thus

$$V_L = IR_L = \frac{R_L(x/L)V_s}{R_p(x/L)[1 - (x/L)] + R_L}$$

$$= \frac{(x/L)V_s}{(R_p/R_L)(x/L)[1 - (x/L)] + 1} \qquad [6]$$

The relationship between $V_L$ and $x$ is non-linear. Hence the effect of the loading is to give a non-linearity error. This error is the difference between the open circuit potential difference between the output terminals and the potential difference when the load is between the terminals. Thus

non-linearity error

$$= E_{Th} - V_L$$

$$= (x/L)V_s\left[1 - \frac{1}{(R_p/R_L)(x/L)[1 - (x/L)] + 1}\right] \qquad [7]$$

This can be rearranged to give

$$\text{error} = (x/L)V_s\left[\frac{(R_p/R_L)(x/L)[1 - (x/L)] + 1 - 1}{(R_p/R_L)(x/L)[1 - (x/L)] + 1}\right]$$

If $R_L \gg R_p$ the denominator of the equation becomes approximately 1 and so the equation approximates to

$$\text{error} = V_s(R_p/R_L)[(x/L)^2 - (x/L)^3] \qquad [8]$$

The maximum value of this error occurs when

$$\frac{d(\text{error})}{dx} = V_s(R_p/R_L)[(2x/L^2) - (3x^2/L^3)] = 0$$

Hence $2x/L^2 = 3x^2/L^3$ and $(x/L) = 2/3$. Substituting this value into equation [8], the maximum error is given by

$$\text{maximum non-linearity error} = V_s(R_p/R_L)[(2/3)^2 - (2/3)^3]$$

$$= 0.148\ V_s(R_p/R_L) \qquad [9]$$

**Example 2**

What is the maximum non-linearity error produced when a load of $100\,\text{k}\Omega$ is connected across a potentiometer of track resistance $100\,\Omega$ if the voltage applied between the ends of the potentiometer track is $2\,\text{V}$?

*Answer*

Using equation [9]

$$\text{maximum error} = 0.148V_s(R_p/R_L)$$

$$= 0.148 \times 2(100/100 \times 10^3) = 0.3\,\text{mV}$$

**Noise**

The term *noise* is generally used for the unwanted signals that may be picked up by a measurement system and interfere with the signal being measured, thus giving rise to random errors. There are two types of noise, *interference* which is due to the interaction between external electrical and magnetic fields and a measurement system circuit, e.g., the circuit picking up interference from nearby mains power circuits, and *random noise* which is due to the random motion of electrons and other charge carriers in components and is a characteristic of the basic physical properties of components in the system.

**Random noise**

Random noise can arise in a number of ways.

1  *Thermal noise*  This is sometimes referred to as *Johnson noise* and is the most common form of noise in a conductor. Other than at absolute zero all the free electrons and other charge carriers in resistors and semiconductors are in completely random motion (Fig. 3.5) and these random motions give rise to a multitude of random currents. This random motion occurs because the material is at some temperature. On average there will be as many electrons moving in one direction as another. However, the numbers will fluctuate about this average and thus even in the absence of an externally applied potential difference there will be a fluctuating current. When there is an externally applied potential difference there is just a drift in the direction of the potential

**Fig. 3.5**  Thermal noise

difference superimposed on the random motion. The fluctuating current arising from the random motion of the electrons is what is termed thermal noise, since the higher the temperature the greater the amount of random motion. The noise is spread over an infinite range of frequencies and is thus referred to as *white noise* (white light is a wide range of frequencies). The root-mean-square (r.m.s.) noise voltage for a bandwidth of $f_1$ to $f_2$ is

$$\sqrt{[4kRT(f_2 - f_1)]} \qquad [10]$$

where $k$ is Boltzmann's constant ($1.38 \times 10^{-23}$ J/K), $R$ the resistance and $T$ the absolute temperature. Thus a wideband amplifier will generate more noise than a narrowband one, a high resistance will generate more noise than a lower one and the higher the temperature the more the noise. The thermal noise is independent of the actual current flowing and occurs regardless of whether it is d.c. or a.c.

2   *Shot noise*  In many electronic components there is a movement of charge carriers across potential barriers, e.g., in a p–n junction. The charge carriers do not flow across the barrier in a constant rate of flow but have a random element superimposed on the motion. There are thus random fluctuations in the rate at which charge carriers diffuse across potential barriers, such fluctuations being known as shot noise. The r.m.s. noise voltage is, for an absolute temperature $T$ and bandwidth of frequency $f_1$ to $f_2$,

$$\sqrt{[2kTr_d(f_2 - f_1)]} \qquad [11]$$

where $k$ is Boltzmann's constant and $r_d$ the differential diode resistance, this being $kT/qI$ where $q$ is the charge on the electron and $I$ the d.c. current in the junction.

3   *Flicker noise*  This is noise which occurs as a result of the flow of charge carriers in a discontinuous medium, e.g., a carbon composite resistor, a diode, transistor or thermistor. We can consider the discontinuities in the medium to result in fluctuations in the velocity of the charge carriers as they drift in the direction imposed on them by a potential difference. The r.m.s. noise voltage is approximately inversely proportional to the frequency and hence is a low frequency noise. For this reason it is often referred to as *pink noise* (pink is the low frequency end of the spectrum of visible light). For a transistor, this noise only becomes apparent below about 1 kHz.

**Noise generators**

A resistor which is generating noise can be considered to be

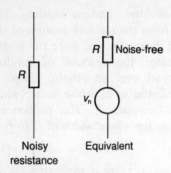

**Fig. 3.6** Noisy resistance

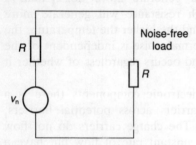

**Fig. 3.7** Maximum power transfer

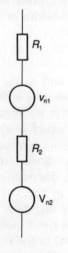

**Fig. 3.8** Noisy resistors in series

equivalent to a noise-free resistance in series with a noise generator (Fig. 3.6). In the case where the only noise is thermal noise the generator will have an r.m.s. voltage $v_n$ given by equation [10] as

$$v_n = \sqrt{[4kRT(f_2 - f_1)]}$$

The maximum power that can be produced in a load is when the load has the same resistance as the source, i.e., they are matched (Fig. 3.7). The voltage drop across the generator resistance $R$ will be the same as that across the load resistance $R$ and so the voltage across the load is

$$\text{voltage across load} = \tfrac{1}{2}\sqrt{[4kRT(f_2 - f_1)]}$$

Thus the maximum noise power that a thermal noise source can supply to a load is

$$\text{max. power} = \frac{\{\tfrac{1}{2}\sqrt{[4kRT(f_2 - f_1)]}\}^2}{R} = kT(f_2 - f_1) \qquad [12]$$

When we have two noisy resistors in series then the equivalent circuit will be as shown in Fig. 3.8 and will involve two r.m.s. voltage generators in series. Each generator will produce completely random voltages, varying in frequency, phase and instantaneous magnitude. Thus the total voltage cannot be obtained just by adding arithmetically the two generator voltages. The total voltage will be that which would have occurred with a single resistor of resistance $(R_1 + R_2)$. Thus the voltage is, by equation [10],

$$v_n = \sqrt{[4k(R_1 + R_2)T(f_2 - f_1)]}$$

Thus

$$v_n^2 = [4kR_1T(f_2 - f_1)] + [4kR_2T(f_2 - f_1)]$$
$$= v_{n1}^2 + v_{n2}^2$$

The resultant noise voltage squared is the sum of the squares of the individual noise voltages.

**Example 3**

What will be (*a*) the thermal noise voltage generated and (*b*) the maximum thermal noise power transferred to a load for a 1 MΩ resistor at a temperature of 20 °C if the effective noise bandwidth is 10 MHz?

*Answer*

(*a*) The noise voltage is given by equation [10] as

$$v_{rms} = \sqrt{[4kRT(f_2 - f_1)]}$$

$$= \sqrt{[4 \times 1.38 \times 10^{-23} \times 10^6 \times 293 \times 10 \times 10^6]}$$

$$= 0.40\,\text{mV}$$

(b) The maximum thermal noise power is given by equation [12] as

$$\text{power} = kT(f_2 - f_1)$$

$$= 1.38 \times 10^{-23} \times 293 \times 10 \times 10^6 = 4.0 \times 10^{-14}\,\text{W}$$

**Signal-to-noise ratio**

A term which is used to quantify the amount of noise present with any given signal is the *signal-to-noise* ratio. This is the ratio of the signal power to the noise power.

$$\text{S/N ratio} = \frac{\text{signal power}}{\text{noise power}}$$

The ratio is usually expressed in decibels as

$$\text{S/N ratio} = 10\lg\left(\frac{\text{signal power}}{\text{noise power}}\right) \qquad [13]$$

Since power is $V^2/R$ then if $V_s$ is the signal voltage, the signal power is $V_s^2/R$ and if $V_n$ is the noise voltage across the same resistor then the noise power is $V_n^2/R$. Thus

$$\text{S/N ratio} = 10\lg\left(\frac{V_s}{V_n}\right)^2 = 20\lg\left(\frac{V_s}{V_n}\right) \qquad [14]$$

Because noise signals are random signals superimposed on the measurement signal, taking the average value of a d.c. signal over a period of time can be used to reduce the effect of noise and so enhance the signal-to-noise ratio. The signal-to-noise ratio is increased in proportion to the square root of the time over which the average is taken. For a repetitive signal averaging can also be used to enhance the signal-to-noise ratio. For the same point in the signal waveform samples are taken for a number of cycles and the average value obtained. For each sample, because the noise is random, the noise signal will be different; sometimes negative, sometimes positive. The result of the averaging the samples is thus to reduce the effect of the noise and so improve the signal-to-noise ratio for that point in the waveform. This averaging process can be repeated for a number of points in the waveform and the signal reconstructed. This technique of averaging is often carried out as a result of the software used with intelligent instruments, i.e., instruments incorporating microprocessors.

### Example 4

Determine the signal-to-noise ratio for a signal input to a system if the voltage of the signal across the input terminals is $4\,\text{mV}$ r.m.s. and the noise voltage $1\,\mu\text{V}$ r.m.s.

*Answer*

Using equation [14]

$$\text{S/N ratio} = 20 \lg \left( \frac{V_s}{V_n} \right) = 20 \lg \left( \frac{4 \times 10^{-3}}{1 \times 10^{-6}} \right) = 72\,\text{dB}$$

**Noise factor**

An instrument may have an input of a signal containing noise and then itself introduce further noise. The output from the instrument thus contains noise resulting from that in the input signal and that introduced by the instrument. The *noise factor* or *noise figure* (NF) is a measure of the amount of noise introduced by the instrument and expresses the degradation of the S/N ratio introduced by the instrument. The noise factor is given by

$$\text{NF} = \frac{(\text{S/N})_{in}}{(\text{S/N})_{out}} \tag{15}$$

or if expressed in decibels as the noise figure

$$\text{NF} = 10 \lg \frac{(\text{S/N})_{in}}{(\text{S/N})_{out}} \tag{16}$$

where $(\text{S/N})_{in}$ is the signal-to-noise ratio of the input and $(\text{S/N})_{out}$ that of the output.

**Interference**

The different types of interference are generally classified according to the physical phenomena responsible for their generation, namely:

1 *Inductive coupling* (or *electromagnetic coupling* or *magnetic coupling*) A changing current in a nearby circuit produces a changing magnetic field and this, as a result of electromagnetic induction, induces e.m.f.s within the measurement system. Common sources of such interference are the currents in power cables in nearby walls and the cables to the instrument itself, and abruptly changing currents that occur in the operation of relays and motors.

2 *Capacitative coupling* The earth and such items as nearby power cables are separated by just a dielectric, generally air, from conductors in the measurement system. The conductor–dielectric–earth and conductor–dielectric–power cable systems are capacitors. A change in the voltage applied to one of the plates of such capacitors affects the voltage on the other. These capacitors thus couple the measurement system conductors to other systems and therefore signals in those systems pass to the measurement system as interference. Capacitative interference signals from nearby mains power cables and appliances will have

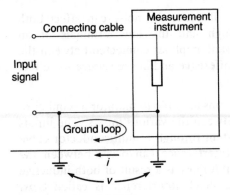

**Fig. 3.9**   Ground loop

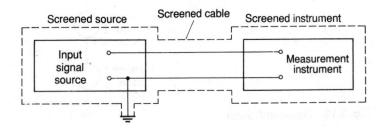

**Fig. 3.10**   Reduction of interference
by twisted pairs

the mains frequency 50 Hz or 60 Hz depending on the country concerned.

3   *Multiple earths*   If the measurement system has more than one connection to earth then an interference current through the measurement system will be produced if all the earth points are not precisely at the same potential. The resulting circuit is known as a *ground loop* (Fig. 3.9).

Interference can be reduced by:

1   *Minimizing cable lengths and loops*   Minimizing cable lengths can reduce the amount of capacitive coupling. The production of e.m.f.s in cables due to electromagnetic induction can be minimized if the loop area formed by cables is minimized. This is because the induced e.m.f. is proportional to the rate of change of the magnetic flux linked by the loop and thus depends on the loop area. A particularly important loop to avoid is one between a cable and the ground.

2   *Twisted pairs of wires*   Currents flowing through a pair of parallel wires will produce magnetic fields which interact with each other and so produce interference. Such magnetic fields can, however, be cancelled if the wires are twisted. This is because we can consider the twisted wire to be essentially a sequence of current carrying loops, with the currents in successive loops in opposite directions (Fig. 3.10). Hence the magnetic fields produced within each loop will be in opposite directions and so the e.m.f.s induced in that portion of wire within each loop will be in opposite directions in successive loops.

3   *Electrostatic screening*   Capacitance coupling can be eliminated by completely enclosing circuits in an earthed metal screen (Fig. 3.11). Coaxial cable gives screening of connections between elements of measurement systems; however, the cable should only be earthed at one end so that multiple earths are avoided.

4   *Single earth*   Multiple earthing should be avoided by only having one earthing point.

5   *Filters*   Filters can be used which transmit the measurement signal but block the interference signal.

**Fig. 3.11**   Screened system

**6** *Differential amplifier* Because the noise can affect both the inputs to the high and the low terminals of an instrument, a differential amplifier connected between the two will amplify the difference and hence reduce the effect of the noise.

Some measurement systems are able to monitor a number of quantities by switching between the various inputs. The inputs are often fed into the system through a multiconductor cable or a ribbon conductor. Interference can occur between the signals in these various input lines as a result of both inductive and capacitive coupling. Such interference is called *cross talk*. The interference may be reduced by increasing the spacing between the conductors, screening the most affected circuits, or with a ribbon conductor by interspersing signal-carrying conductors with earthed conductors.

**Noise rejection**

Noise may arise within the signal source and is referred to as *normal mode noise* or may occur between the earth terminal of the instrument and its lower potential terminal, in which case it is referred to as *common mode noise*. Figure 3.12 shows the equivalent circuit for an instrument being used to make a measurement when there is both normal mode noise and common mode noise. The normal mode noise is represented by a voltage generator in series with the signal being measured while the common mode noise is a generator connected between the lower potential terminal of the instrument and earth.

To the measurement system normal mode noise is indistinguishable from the signal being measured. Such noise may be caused by interference due to inductive and capacitive coupling with external power cables. The ability of a system to suppress common mode voltages, i.e., reject normal mode noise, is called the *normal mode rejection ratio* (NMRR). This can be defined, in decibels, as

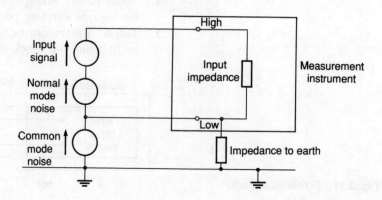

**Fig. 3.12** Equivalent circuit

$$NMRR = 20 \lg \left( \frac{V_n}{V_e} \right) \qquad [17]$$

where $V_n$ is the peak value of the normal mode noise and $V_e$ is the peak value of the error it produces in the measurement at a particular frequency. An alternative way of describing the *NMRR* is in terms of the peak value of the normal mode noise which would not produce an error greater than some specified error value. Normal mode noise may be reduced by using filters to isolate the signal being measured and block the interference.

The term *common mode noise* is used to describe the noise occurring between the earth terminal of a measurement system and its lower potential terminal. It results from earth loop currents which may arise because there are multiple earths, between which there is a potential difference, or from electromagnetic induction inducing currents in the earth loop. The ability of a measurement system to prevent common mode noise introducing an error in the measurement reading is called the *common mode rejection ratio* (*CMRR*).

$$CMRR = 20 \lg \left( \frac{V_{cm}}{V_e} \right) \qquad [18]$$

where $V_{cm}$ is the peak value of the common mode noise and $V_e$ the peak value of the error it produces in the measurement at a particular frequency.

The *CMRR* can be improved by having only a single earthing point, using differential amplifiers, isolating the instrument input with respect to earth (this is called *floating*) and using shielding. Figure 3.13 shows a differential amplifier used with shielding of the input cable and the amplifier. The amplifier is floating since neither of its input terminals is connected to earth.

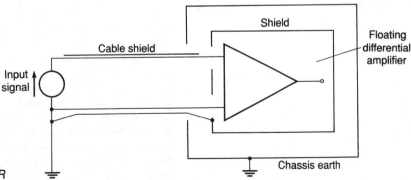

**Fig. 3.13**  Improving the *CMRR*

## Other sources of error

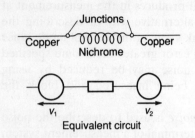

Fig. 3.14 Dissimilar metal junctions

Other common sources of error with electrical measurements are the failure to take account of the resistance of connecting leads between a resistance element and the measurement system and e.m.f.s produced as a result of thermoelectricity. An e.m.f. is produced at the junction of two different metals, the size of the e.m.f. depending on the temperature of the junction. This is the principle of the thermocouple. Thus connecting a wire-wound resistor into a circuit where the wires are of a different metal can produce a thermal e.m.f. There will be two dissimilar metal junctions, one at each end of the resistor (Fig. 3.14). If both junctions are at the same temperature, the two thermoelectric e.m.f.s will cancel; however, if there is a temperature difference between the two ends this will not be the case.

## Intelligent instruments

An instrument is said to be *intelligent* if it includes a microprocessor: with no such element the instrument is said to be *dumb*. With an intelligent instrument the measurements are made in the same way as with a dumb instrument but the difference is that the intelligent instrument is able to carry out calculations, combine measurements from a number of sources, make decisions based on a sequence or number of measurements, manipulate information and initiate actions.

Intelligent instruments can be used to reduce both random and systematic errors. Thus, as already discussed earlier in this chapter, they can use averaging to reduce the effects of random noise. If there is a loading effect the intelligent instrument can readily correct the measured value for the loading error. They can also monitor the temperature, or other environmental conditions, and correct the measured value for changes in such conditions. Other corrections may be carried out, e.g., for drift and non-linearity, and the calibration can be automatically checked.

## Problems

1  What is the percentage error of a voltmeter with a resistance of 1 MΩ when used to measure a circuit voltage if the circuit has a Thévenin equivalent resistance of (*a*) 1 kΩ, (*b*) 50 kΩ?
2  What resistance should an ammeter have to measure, with an error of no more than 5%, the current in a circuit having a resistance of 300 Ω?
3  What is the maximum non-linearity error as a percentage of the voltage input that is produced when a load of 500 Ω is connected across a potentiometer of track resistance 100 Ω?
4  What is the percentage error in the voltage reading given by a voltmeter of resistance 10 kΩ when used to measure the voltage between the terminals of the circuit shown in Fig. 3.15?
5  What will be the thermal noise voltage generated for a 100 kΩ

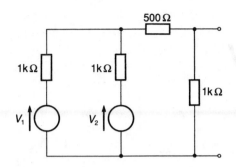

**Fig. 3.15**   Problem 4

resistor at a temperature of 18 °C if the effective noise bandwidth is 10 MHz?

6   Determine for a bandwidth of 100 kHz the thermal noise voltage produced by (*a*) a 20 kΩ resistor, (*b*) a 50 kΩ resistor and (*c*) the two resistors in series if they are at a temperature of 17 °C?

7   Explain how thermal noise arises and why it is termed 'white noise'. How will such noise change for a resistor if the temperature of the resistor is increased from 0 to 25 °C?

8   Determine the signal-to-noise ratio for a signal input to a system if the voltage of the signal across the input terminals is 10 mV r.m.s. and the noise voltage 1 μV r.m.s.

9   What is the signal-to-noise ratio for an amplifier which has a noise signal of 20 nV/√Hz bandwidth r.m.s. when it is operated with a bandwidth of 1 kHz to 10 kHz with a signal of 20 mV r.m.s.?

10   Explain how (*a*) capacitive coupling and (*b*) inductive coupling can cause interference to be introduced into a measurement system and steps that can be taken to reduce such interference.

11   Explain what is meant by a ground loop. Explain how ground loops can exist with a system even when there is only one point in the system earthed.

12   What is the difference between normal mode noise and common mode noise? Explain how it can be reduced.

# 4 Reliability

## Introduction

If you buy a TV set or a car a vital factor in determining which one to buy is the reliability, i.e., how long you can rely on it performing the required function without breaking down or not performing to specification. *Reliability* is defined as being the chance that a product will operate to the specified level of performance for a specified period under specified environmental conditions. Thus the TV set might have a guarantee for two years, indicating that the manufacturer is stating that the customer can expect reliable performance for that time. In choosing measurement systems a vital factor in the choice is the reliability. For how long can the system be expected to perform to specification? What is the chance of the system failing to perform to specification? Reliability is an essential aspect of quality assurance.

## Reliability and chance

The *reliability* of a product is defined as the chance that it will operate to a specified level of performance for a specified period under specified environmental conditions. The product concerned might be just components used in a measurement system, e.g., electrical components such as resistors, mechanical components such as suspension wires, modules such as amplifiers, or the system as a whole. The specified level of performance might be an accuracy of ±1%; performance outside this would be considered to be a failure to meet the specified performance specification. The environmental conditions for which the reliability is specified might be temperatures in the range 10–30 °C. When the product is used outside those limits then the manufacturer's guarantee regarding the reliability of the product is not valid.

Consider the reliability of an electric light bulb. It is not possible to say that every bulb will have a life of, say, 1000 hours and then all will burn out. This is because the materials that are used in the manufacture of the bulb are not perfect, but fall within certain tolerances; also the conditions under which the bulbs are used will not all be identical. There will thus be random variations between bulbs. Because of this the operating life of bulbs cannot be specified in absolute terms but only predicted on the basis of chance or probability. Thus there might be a very low probability that a bulb will burn out before 900 hours, a higher probability for between 900 and 1000 hours, and a much higher probability with more than 1000 hours.

*Chance*, or *probability*, is the frequency in the long run with which an event occurs. If a coin is dropped it might land heads or tails and it is not possible to predict which will occur (assuming no bias in the coin or the way it is dropped). If the coin is dropped 10 times then we might find it lands heads uppermost 3 times, i.e. the frequency of heads is 3/10 = 0.3. If the coin is dropped 20 times then we might find it lands heads uppermost 12 times, a frequency of 12/20 = 0.6. If the coin is dropped 100 times then we might find it is heads uppermost 47 times, a frequency of 0.47. If the coin is dropped 1000 times then we might find it is heads uppermost 501 times, a frequency of 0.51. The frequency with which heads occurs settles down to an almost constant figure of 0.50 as the number of times the coin is dropped gets larger. In the long run, the frequency can be said to be 0.50. Thus the probability of heads can be said to be 0.50.

The probability of heads uppermost with a dropped coin is 0.50 or $\frac{1}{2}$. This is because each time a coin is dropped there are two possible ways the coin can land and thus the chance of it landing one way is 1 in 2 or $\frac{1}{2}$. The chance of the coin landing with either heads or tails uppermost is 2 in 2 or 1. A chance of 1 is a certainty. A chance of 0 means it will never happen.

If we start off with a large number of items $N_0$ and after some time $t$ there are $N$ left which still meet specification, then we state that the reliability at time $t$ is

$$\text{reliability} = \frac{N}{N_0} \qquad [1]$$

i.e., the chance after time $t$ of finding an item meeting specification is $N/N_0$.

The *unreliability* of an item is the chance that it will fail to operate to specification for a specified period under specified environmental conditions. Thus if we start off with a large number of items $N_0$ and after some time $t$ there are $N$ left which still meet specification, then the number failing in that time is $(N_0 - N)$. Hence

$$\text{unreliability} = \frac{N_0 - N}{N_0} \qquad [2]$$

Hence, since this equation can be written as $1 - (N/N_0)$ using equation [1],

$$\text{unreliability} = 1 - \text{reliability} \qquad [3]$$

**Example 1**

What is the reliability of electric light bulbs for a time of 1000 hours if for a batch of 1000 it is found that 50 fail in that time?

*Answer*

Using equation [1] then

$$\text{reliability} = \frac{N}{N_0} = \frac{950}{1000} = 0.95$$

We can state that there is a 95% chance that an electric light bulb will still operate after 1000 hours.

**Failure**

*Failure* can be defined as being when a system fails to perform to the specified level of performance. The failure may be complete in that the system can no longer be used or partial in that the system can still be used but has just gone outside the specified limits. While reliability is just the probability of not failing, a useful measure of reliability is in terms of the mean time between failures, the failure rate or the mean time to failure.

Suppose we find for an item that $N_f$ failures occur during a time $t$, the item being repaired each time it fails. Then the mean time between failures is $t/N_f$. If $N$ items are tested for a time $t$ with failed items being repaired and put back into service and during that time there are $N_f$ failures, then the *mean time between failures (MTBF)* is

$$MTBF = \frac{N_t}{N_f} \qquad [4]$$

The term *failure rate* $\lambda$ can be used for the average number of failures per unit time for an item, the item being repaired each time it fails. Thus if over a time $t$ there are $N_f$ failures then the failure rate is $N_f/t$. If we have $N$ items being tested for a time $t$, with failed items being repaired and put back into service, and $N_f$ failures occur then the failure rate per item is $N_f/Nt$:

$$\lambda = \frac{N_f}{Nt} \qquad [5]$$

Hence

$$\lambda = \frac{1}{MTBF} \qquad [6]$$

**Table 4.1** Failure rates and *MTBF*

| Component | Failure rate $\times 10^{-6}$ per hour | MTBF $10^6$ hours |
|---|---|---|
| Filament lamp | 5 | 0.2 |
| Electrolytic capacitor | 1.5 | 0.7 |
| Paper capacitor | 1 | 1 |
| Silicon transistor (>1 W) | 0.8 | 1.3 |
| Carbon resistor | 0.5 | 2 |
| Wire wound resistor | 0.1 | 10 |
| Plastic film capacitor | 0.1 | 10 |
| Silicon transistor (<1 W) | 0.08 | 12.5 |
| Semiconductor diode | 0.05 | 20 |
| Soldered connection | 0.01 | 100 |
| Wrapped connection | 0.001 | 1000 |

Table 4.1 shows some typical values of failure rates and mean time between failures for components.

A failure rate of $5 \times 10^{-6}$ per hour for the filament lamp does not mean that if $1/(5 \times 10^{-6}) = 2 \times 10^5$ lamps are observed for one hour that exactly one will fail. The failure rate is the average value for large numbers of items and large numbers of failures. The failure rate is thus only an estimate of the likely failure rate.

The *mean time to failure* (*MTTF*) is the average time to failure for a number of samples of a product, it being assumed that it is impossible or uneconomic to repair them. Thus if $N$ items are tested and the time to failure for each is $t_1$, $t_2$, $t_3$, ..., $t_N$, then the average time to failure is

$$MTTF = \frac{t_1 + t_2 + t_3 + \ldots + t_N}{N} \qquad [7]$$

**Example 2**

A measurement system is found over a period of 1000 hours to have failed 10 times, being repaired at each failure. What is (*a*) the failure rate, and (*b*) the mean time between failures?

*Answer*

(*a*) The failure rate is the average number of failures per unit time and thus is $10/1000 = 0.01$ per hour.
(*b*) The mean time between failures is $1000/10 = 100$ hours.

**Availability**

The mean time between failures gives the average time that an item is usable between failures. When it fails it is repaired. This takes time. We can measure the *mean time to repair*

(*MTTR*). The fraction of the total time for which the item is in a usable state is thus

$$\text{availability} = \frac{MTBF}{MTBF + MTTR} \qquad [8]$$

Since $MTBF = 1/\lambda$, where $\lambda$ is the failure rate, this equation can also be written as

$$\text{availability} = \frac{1}{1 + \lambda \times MTTR} \qquad [9]$$

### Example 3

A system has a mean time between failure of 6000 hours and a mean time to repair of 5 hours. What is the availability?

*Answer*

Using equation [8],

$$\text{availability} = \frac{MTBF}{MTBF + MTTR} = \frac{6000}{6000 + 5} = 0.9992$$

## Failure rate and time

The following might be considered a typical set of times to failure, in hours, for a product:

1, 1, 1, 2, 2, 3, 20, 40, 50, 70, 100, 104, 105, 105, 105

In the first hour the failure rate is 3 per hour, dropping to 2 per hour in the second hour and 1 per hour in the third hour. From 3 to 100 hours the failure rate is on average 5/97 per hour, being reasonable constant at about 1/20 per hour. Above 100 hours the failure rate increases. If the pattern of failure for a product is studied it is often found that the failure rate varies with time in the way shown in Fig. 4.1. Because of its shape the graph is referred to as the 'bath tub curve'. Such a graph shows three distinct phases:

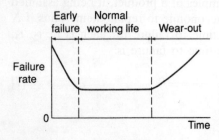

**Fig. 4.1** The 'bath tub curve'

1　Early failure due to manufacturing faults, sub-standard components, material imperfections, faulty assembly, bad connections, etc. Where such failures are likely to occur, manufacturers 'burn in' the products, i.e., run them for this period of time, so that any that will fail due to such faults will do so before they are sent to a customer.
2　Normal working life, where the failure rate is virtually constant and the result of purely random events.
3　Wear-out failure, where the failure rate increases due to components wearing out.

## The exponential law of reliability

Consider a situation where a product has a constant failure rate $\lambda$ and what we are concerned with is how the number of items which have had no failures varies with time. We can simulate such a situation by taking, say, 100 coins and supposing that when we drop them, those falling heads uppermost have failed. When a coin 'fails' we withdraw it from the pile. Thus after the first drop we might have 50 failures and so be left with 50 'unfailed' coins. The chance of 'failure', i.e., obtaining heads uppermost, with coins is $\frac{1}{2}$. Then the coins are dropped again and again we find the number of 'failures' and withdraw them. Because the failure rate is constant we would expect half to fail and so there would be about 25 coins failed and 25 unfailed. This can be repeated a number of times. Predicting the number of unfailed coins on the basis of a constant chance of failure of $\frac{1}{2}$ leads to the number of unfailed coins (in rounded numbers) decreasing as follows:

100   50   25   13   6   3

The way in which the number of coins decreases with the number of drops is an exponential.

Now consider the failure of products when there is a constant chance of failure, as with the coins. Suppose that initially, at time $t = 0$, there are $N_0$ items. As time increases so the number of unfailed items will decrease. Consider that by time $t$ the number of items has decreased to $N$. After a further very small interval of time $\delta t$ the number drops to $(N - \delta N)$. The number failing is thus $-\delta N$. Thus the failure rate is

$$\lambda = \frac{-\delta N}{N \delta t}$$

In the limit as $\delta t$ tends to zero, then the expression becomes

$$\lambda = -\frac{1}{N}\frac{dN}{dt}$$

On rearranging this, and integrating between the values at zero time and $t$,

$$\int_{N_0}^{N} \frac{dN}{N} = -\int_{0}^{t} \lambda \, dt$$

$$\ln N - \ln N_0 = -\lambda t$$

and so the way in which the number of unfailed items varies with time is given by

$$N = N_0 \exp(-\lambda t) \qquad [10]$$

Thus the number of unfailed items varies exponentially with time. The reliability is $N/N_0$, hence

$$\text{reliability} = \exp(-\lambda t) \qquad [11]$$

The reliability thus decreases exponentially with time. The

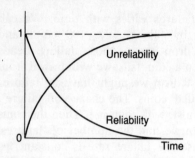

**Fig. 4.2** Variation of reliability and unreliability with time

total number of failures in the time $t$ is $(N_0 - N)$. Hence

$$\text{number of failures} = N_0 - N_0\exp(-\lambda t)$$

The unreliability is (number of failures/$N_0$). Hence

$$\text{unreliability} = 1 - \exp(-\lambda t) \qquad [12]$$

Figure 4.2 shows graphs of the reliability and unreliability variations with time, i.e., equations [11] and [12].

**Example 4**

An instrument has an estimated mean time between failures of 1000 hours. What is the chance of it working successfully after 500 hours?

*Answer*

The average failure rate is, using equation [6],

$$\lambda = \frac{1}{MTBF} = \frac{1}{1000} = 0.001 \text{ per hour}$$

Assuming that the failure rate is constant, equation [11] gives

$$\text{reliability} = \exp(-\lambda t) = \exp(-0.001 \times 500) = 0.61$$

**Reliability of a system**

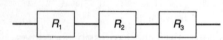

**Fig. 4.3** Elements in series

With many systems containing a number of components, the entire system fails if just one component fails. Such a situation is like a number of lamps wired in series. When one lamp fails, all the lamps go out. Such systems can indeed be referred to as consisting of elements in series. Consider a system consisting of a number of elements in series (Fig. 4.3) with each of the elements having its own, independent, reliability. The entire system fails if one of the elements fails. To establish the reliability of the entire system we need to find out the chance of one of the system elements failing.

The chance of one coin landing with heads uppermost is 1/2. The chance of one of the three coins landing with heads uppermost is $7 \times (1/2) \times (1/2) \times (1/2)$ or 7/8. This can be worked out by considering all the possible ways the coins can land. Only one of the ways has no heads. Thus the chance of there being no failure is 1/8, i.e., $1/2 \times 1/2 \times 1/2$.

H T T  H H T  H H H  H T H

T T T  T H T  T H H  T T H

Hence the reliability $R_s$ of a three-element system is

$$R_s = R_1 \times R_2 \times R_3 \qquad [13]$$

where $R_1$, $R_2$ and $R_3$ are the reliabilities of the elements.

The reliability of the system $R_s$ can be expressed in terms of the system failure rate $\lambda_s$ as

$$R_s = \exp(-\lambda_s t)$$

Hence since $R_1 = \exp(-\lambda_1 t)$, $R_2 = \exp(-\lambda_2 t)$ and $R_3 = \exp(-\lambda_3 t)$ then

$$\exp(-\lambda_s) = \exp(-\lambda_1 t) \times \exp(-\lambda_2 t) \times \exp(-\lambda_3 t)$$
$$= \exp-(\lambda_1 + \lambda_2 + \lambda_3)t$$

Hence

$$\lambda_s = \lambda_1 + \lambda_2 + \lambda_3 \qquad [14]$$

The failure rate of the system with series elements is the sum of the failure rate of the constituent elements.

### Example 5

A system contains 20 integrated circuits, each having a failure rate of $1 \times 10^{-6}$ per hour. What is the chance of the system operating for 10 000 hours without failure if the failure of any one of the circuits will result in failure of the system?

*Answer*

The system consists of 20 elements in series. Thus the failure rate for the system is the sum of the failure rates of each circuit (equation [14]) and so is $20 \times 1 \times 10^{-6}$ per hour. The chance of the system operating for 10 000 hours without failure is the reliability at that time. This is given by equation [11] as

$$\text{reliability} = \exp(-\lambda t) = \exp(-20 \times 10^{-6} \times 10\,000) = 0.82$$

### Elements in parallel

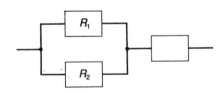

**Fig. 4.4** Elements in parallel

If a number of lamps are connected in parallel then the failure of one of the lamps does not result in the system failing to provide light, unlike a system of lamps wired in series. A parallel connection of elements in any system means that there is an alternative element in parallel with it (Fig. 4.4) and the system can only then fail if all the parallel elements fail. Where it is essential that a system should not fail, e.g., power systems in an aircraft, then there can be doubly installed systems so that if one system fails there is still another system in a working condition. The connection of systems in parallel is said to introduce *redundancy*.

For elements in parallel the chance of failure is the chance that both items will simultaneously fail. The chance of this is like dropping two coins and considering what the chance is of both landing heads uppermost. The different ways the two coins can land are

H T   H H   T T   T H

Thus the chance of two heads is 1/4, i.e., $1/2 \times 1/2$.

The chance of failure of an element is its unreliability. The unreliability of one of the parallel elements in Fig. 4.4 is $(1 - R_1)$, the unreliability of the other $(1 - R_2)$. The chance that both elements will fail is the product of the individual probabilities because the probability of two independent events occurring is the product of the individual probabilities. Hence the unreliability of both elements failing is

$$\text{unreliability of system} = (1 - R_1)(1 - R_2) \qquad [15]$$

The reliability of the arrangement is

$$\text{reliability} = 1 - (1 - R_1)(1 - R_2)$$
$$= R_1 + R_2 - R_1 R_2 \qquad [16]$$

The above discussion has assumed that both parallel elements are simultaneously operating all the time, i.e., both are on-line. An alternative is to have only one of the elements operating and the other on stand-by to be switched on immediately the other one fails.

### Example 6

A system has for safety reasons two parallel power systems. if each of the power systems has a reliability of 0.90 after 10000 hours of operation, what is the reliability of the parallel arrangement?

*Answer*

Using equation [16]

$$\text{reliability} = R_1 + R_2 - R_1 R_2$$
$$= 0.90 + 0.90 - 0.90 \times 0.90 = 0.99$$

The parallel arrangement has increased the overall reliability.

**Problems**

1  What is the reliability after a million hours for carbon film resistors if after testing 10000 for that time four failures were found?

2  If the reliability of a component is 0.90, what is its unreliability?

3  A measurement system is found over a period of 1000 hours to have failed four times, being repaired at each failure. What is (a) the failure rate, and (b) the mean time between failures?

4  The failure rate of a component is $0.04 \times 10^{-6}$ per hour. What is the mean time between failures?

5  A system has a mean time between failure of 2000 hours and a mean time to repair of 4 hours. What is the availability?

6  A system has an estimated mean time between failures of 10000 hours. What is the chance of it working successfully after (a) 100 hours, (b) 5000 hours, (c) 10000 hours, (d) 50000 hours?

7  A system has a constant failure rate of $2.0 \times 10^{-6}$ per hour. What will be the reliability after 10000 hours?

8  A system contains 20 resistors, each having a failure rate of $1 \times$

$10^{-8}$ per hour, and 23 soldered joints, each having a failure rate of $10 \times 10^{-8}$ per hour. What is the chance of the system operating for 10 000 hours without failure if the failure of any one of the resistors or joints will result in failure of the system?

9 A system consists of three modules, the individual mean time between failures for the three being 20 000 hours, 50 000 hours and 80 000 hours. What will be the reliability of the system after 10 000 hours if the failure of any one module results in failure of the entire system?

10 An aircraft has a power unit consisting of two identical systems operating simultaneously in parallel. If each has a failure rate of $3 \times 10^{-4}$ hour, what is the reliability after 1000 hours if each can be assumed to have a constant failure rate?

11 A communication system has two identical systems operating simultaneously in parallel. If each has a failure rate of $1 \times 10^{-4}$ per hour, what is the system reliability after 100 hours if each can be assumed to have a constant failure rate?

12 A lighting system consists of four identical bulbs in parallel. If the reliability of a single bulb after 1000 hours is 0.90, what is the reliability of the system?

# 5 Units and standards

Introduction

Measurement involves finding out for an unknown quantity how many multiples of some fixed quantity it is, the fixed quantity being called a *unit*. Measurement thus requires a system of units that is accurate, reliable and easy to use. So that world wide the same units are used with the same values it is necessary to have precise definitions of units and how they can be realized for calibration of measurement systems. Such realizations are referred to as *standards*. In everyday applications, measurements are made using instruments that have been calibrated against a local reference standard which in turn has been compared with a higher-echelon standard which in turn . . ., and so on to calibration against national standards which are set up according to internationally agreed specifications. *Calibration* is the term used for the checking of a measuring system against a standard when the system is in an environment in accordance with that defined for the realization of the standard.

Units

In 1960 the 11th Conférence Générale des Poids et Mesures adopted the Système International d'Unités as an international system of units. This system is referred to as the *SI system*. Later meetings modified the system so that now there are seven basic units: mass in kilograms, length in metres, time in seconds, current in amperes, temperature in degrees kelvin, luminous intensity in candelas and amount of substance in moles. From these basic units other units can be derived. Originally units were based on material standards, e.g., a unit of length based on there being a standard length of metal against which all other length standards were calibrated. However, with the exception of the unit of mass, units are now based on physical phenomena rather than material standards. Thus, for example, length is now based on the distance travelled by light during some time interval. Basing units on physical phenomena enables laboratories anywhere in the world to realize the units without the necessity of calibration against any other standard. The definitions of the basic units are in terms of:

1   *Mass*   The kilogram (kg) is defined as being the mass of an alloy cylinder (90% platinum–10% iridium) of equal height and diameter, held at the International Bureau of Weights and Measures at Sèvres in France. Duplicates of this standard are held in other countries.

2   *Length*   The metre (m) is defined as the length of path travelled by light in a vacuum during a time interval of 1/299 792 458 of a second.

3   *Time*   The second ($s$) is defined as a duration of 9 192 631 770 periods of oscillation of the radiation emitted by the caesium-133 atom under precisely defined conditions of resonance.

4   *Current*   The ampere (A) is defined as that constant current which, if maintained in two straight parallel conductors of infinite length, of negligible circular cross-section, and placed one metre apart in a vacuum, would produce between these conductors a force equal to $2 \times 10^{-7}$ newton per metre of length.

5   *Temperature*   The kelvin (K) is defined so that the temperature at which liquid water, water vapour and ice are in equilibrium (known as the triple point) is 273.16 K.

6   *Luminous intensity*   The candela (cd) is defined as the luminous intensity, in a given direction, of a specified source that emits monochromatic radiation of frequency $540 \times 10^{12}$ Hz and that has a radiant intensity of 1/683 watt per unit steradian (a unit solid angle).

7   *Amount of substance*   The mole (mol) is defined as the amount of a substance which contains as many elementary entities as there are atoms in 0.012 kg of the carbon-12 isotope.

There are also two supplementary units:

1   *Plane angle*   The radian (rad) is the plane angle between two radii of a circle which cuts off on the circumference an arc with a length equal to the radius.

2   *Solid angle*   The steradian (sr) is the solid angle of a cone which, having its vertex in the centre of the sphere, cuts off an area of the surface of the sphere equal to the square of the radius.

Other units are derived from the primary units. Table 5.1 is a list of some commonly used units and their relationship with the primary units.

**Table 5.1** Derived units

| Quantity | Unit name | Unit in terms of primary units |
|---|---|---|
| *Mechanical units* | | |
| Acceleration | metre/second$^2$ | $\mathrm{m\,s^{-2}}$ |
| Angular acceleration | radian/second$^2$ | $\mathrm{rad\,s^{-2}}$ |
| Angular frequency | radian/second | $\mathrm{rad\,s^{-1}}$ |
| Angular velocity | radian/s | $\mathrm{rad\,s^{-1}}$ |
| Area | metre$^2$ | $\mathrm{m^2}$ |
| Density | kilogram/metre$^3$ | $\mathrm{kg\,m^{-3}}$ |
| Energy | joule J | $\mathrm{m^2\,kg\,s^{-2}} = \mathrm{N\,m}$ |
| Force | newton N | $\mathrm{m\,kg\,s^{-2}}$ |
| Frequency | hertz Hz | $\mathrm{s^{-1}}$ |
| Power | watt W | $\mathrm{m^2\,kg\,s^{-3}} = \mathrm{J\,s}$ |
| Pressure | pascal | $\mathrm{kg\,m^{-1}\,s^{-2}} = \mathrm{N\,m^{-2}}$ |
| Speed | metre/second | $\mathrm{m\,s^{-1}}$ |
| Torque | newton metre | $\mathrm{kg\,m^2\,s^{-2}} = \mathrm{N\,m}$ |
| Volume | metre$^3$ | $\mathrm{m^3}$ |
| *Electrical units* | | |
| Capacitance | farad F | $\mathrm{s^4\,A^2\,kg^{-1}\,m^{-2}} = \mathrm{A\,s}$ |
| Conductance | siemen S | $\mathrm{s^3\,A^2\,kg^{-1}\,m^{-2}} = \Omega^{-1}$ |
| Electric charge | coulomb C | $\mathrm{A\,s}$ |
| Electric field strength | volt/metre | $\mathrm{m\,kg\,A^{-1}\,s^{-3}} = \mathrm{V\,m^{-1}}$ |
| Electric potential | volt V | $\mathrm{m^2\,kg\,s^{-3}\,A^{-1}}$ |
| Resistance | ohm $\Omega$ | $\mathrm{m^2\,kg\,A^{-2}\,s^{-3}}$ |
| *Magnetic units* | | |
| Inductance | henry H | $\mathrm{m^2\,kg\,s^{-2}\,A^{-2}}$ |
| Magnetic field strength | ampere/metre | $\mathrm{A\,m^{-1}}$ |
| Magnetic flux | weber Wb | $\mathrm{m^2\,kg\,A^{-1}\,s^{-2}}$ |
| Magnetic flux density | tesla T | $\mathrm{kg\,A^{-1}\,s^{-2}} = \mathrm{Wb\,m^{-2}}$ |

**Electrical standards**

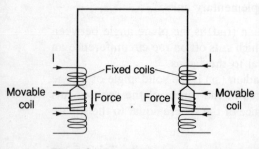

**Fig. 5.1** Current balance principle

The basic electrical unit is the ampere and is defined in terms of the force between two current-carrying conductors. The current balance can be used to realize this unit, Fig. 5.1 being a rough outline of its principle. The force between two current-carrying coils is measured. In principle all electrical measurements could be referred to this instrument. However, this is not very practical. For ease of use in calibration, standards are required which can be stored and just 'pulled off the shelf' when required. Thus, in practice, all national laboratories maintain material standards. Such standards are usually standard cells, standard resistors and standard capacitors. These are generally called the *primary standards* and are the standards used for providing a national calibration service. These material standards are monitored to ensure stability of value by comparison with *reference standards*: the

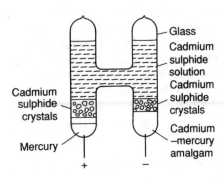

Glass
Cadmium sulphide solution
Cadmium sulphide crystals
Cadmium–mercury amalgam
Cadmium sulphide crystals
Mercury
+      −

**Fig. 5.2**   Weston standard cell

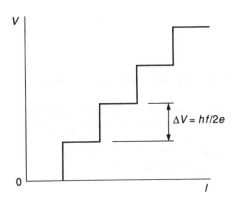

$V$

$\Delta V = hf/2e$

0

$I$

**Fig. 5.3**   Josephson effect

**Traceability**

current balance, the Josephson-junction system, the Campbell mutual inductor and the calculable capacitor.

*Weston mercury–cadmium cells* are used as national primary standards of voltage. Figure 5.2 shows the basic form of such a cell. The nominal value of the e.m.f. from such a cell is 1.018 65 V at 20 °C. Intercomparison of cells can be undertaken to a high degree of precision and show that there is a high degree of stability with respect to each other, of the order of a few parts in $10^7$ per year. Such intercomparisons do not, however, guarantee the absolute value of the voltages. They may all drift by the same amount per year. The *Josephson effect* is now widely used to monitor the absolute value of the voltages. When a thin insulating layer between two superconductors is exposed to microwaves then the voltage–current relationship for the junction shows distinct steps, each voltage step being of size $hf/2e$, where $h$ is Planck's constant, $f$ the frequency of the microwaves and $e$ the charge on the electron (Fig. 5.3). Comparison of junction voltages with standard cells enables standard cells to be maintained absolutely to values with an accuracy of about 3 parts in $10^8$.

*Standard resistors* are used as national primary standards for resistance. Such resistors are wire-wound resistors, the material used for the wire and the methods used to mount the wire being specially chosen to ensure stability. The resistors are immersed in oil. Such resistors have a stability of the order of 1 part in $10^7$ per year. The absolute values of the standard resistors are monitored by means of the *Campbell mutual inductor*. This inductor has a mutual inductance which can be determined from geometric measurements made on the inductor coils. A bridge can be used to determine the resistance of a standard resistor in terms of the mutual inductance. Another form of bridge can also be used to determine the resistance of a standard resistor in terms of the capacitance of a standard capacitor.

*Standard capacitors* as primary standards are constructed from interleaved multiplates suspended in a gaseous dielectric. Silver–mica capacitors tend to be used as secondary standards. The primary standards are monitored by means of a *calculable capacitor*. This is a special capacitor for which the capacitance can be calculated to an absolute accuracy of a few parts in $10^7$.

The standards maintained by national laboratories, e.g., the National Physical Laboratory in Great Britain and the National Bureaux of Standards in the United States, are called *primary standards*. These are then used to calibrate other laboratories' reference standards, such second line of standards being called *secondary standards*. These may be used by

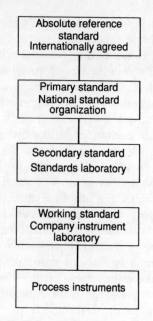

**Fig. 5.4** Simple traceability ladder

calibration centres to carry out calibrations for industry. In a company there may well be such calibrated standards, so-called *working standards*, kept for checking the calibration of instrumentation in day-to-day use. Figure 5.4 shows the traceability ladder relating an instrument being used to make measurements in some process to the standards.

When a working standard has been calibrated by an authorized standards laboratory a calibration certificate is issued. This includes:

1 The identification of the equipment calibrated.
2 The calibration results.
3 The accuracy of the results.
4 Any limitations on the use of the equipment if the calibrated results are to be obtained.
5 The date of the calibration.
6 The authority under which the certificate is issued.

**Problems**

1 Explain what is meant by (*a*) primary units and (*b*) secondary units.
2 Explain the difference between a materials-based standard/unit and a physical phenomena-based standard/unit and the advantages each have.
3 Explain what is meant by traceability and illustrate your answer by indicating a possible traceability ladder for an ammeter in a company or college.

# 6 Analogue meters

Meters may be classified as analogue or digital. With an *analogue meter* the position of the pointer along the instrument scale is related to the size of the quantity being measured. A *digital meter* gives its reading in the form of digits, i.e., a number. This chapter is concerned with analogue meters, and Chapter 7 deals with digital meters. The meters considered with regard to their principles of operation and typical performance are the moving coil meter, electronic multimeter, light spot galvanometer, moving iron meter, hot wire meter, thermocouple meter, electrostatic meter and electro-dynamometer.

**Performance criteria**

Criteria involved in considering the performance of a meter are range, linearity and accuracy. When the meter is to be used for the measurement of alternating current or voltage it is also necessary to consider not only what range of frequencies the meter can be used with but also what aspect of the alternating signal the meter is responding to. With alternating current, meter scale readings can be in terms of the root-mean-square current, the average current or the peak value of the current.

The root-mean-square current is the direct current that would give the same power dissipation in a resistor. Thus for a steady current $I$ to give the same power as the alternating current, whatever its waveform, we must have

$$I^2R = \text{(mean of the sum of the } i^2R \text{ values of the alternating current)}$$

Thus the root-mean-square current is

$$I_{\text{rms}} = \sqrt{\text{(mean of the sum of the values of the squares of the alternating current)}}$$

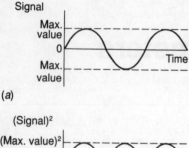

(a)

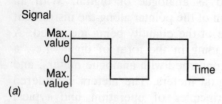

(b)

**Fig. 6.1** Sinusoidal waveform

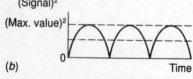

(a)

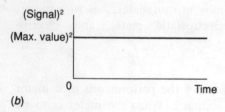

(b)

**Fig. 6.2** Square waveform

Similarly for the root-mean-square voltage

$$V_{rms} = \surd(\text{mean of the sum of the values of the squares of the alternating voltage})$$

For a sinusoidal waveform (Fig. 6.1(a)), squaring the signal results in the graph shown in Fig. 6.1(b) and the mean value of the squared signal is (maximum value)$^2$/2. Hence

$$\text{r.m.s. value of current} = \frac{\text{maximum value of current}}{\surd 2} \qquad [1]$$

$$\text{r.m.s. value of voltage} = \frac{\text{maximum value of voltage}}{\surd 2} \qquad [2]$$

For a square waveform (Fig. 6.2(a)), squaring the signal results in the graph shown in Fig. 6.2(b) and the mean value of the squared signal is (maximum value)$^2$. Hence

$$\text{r.m.s. value of current} = \text{maximum value of current} \qquad [3]$$

$$\text{r.m.s. value of voltage} = \text{maximum value of voltage} \qquad [4]$$

Some meters respond to the mean value of a positive half cycle of the alternating current or voltage. For a sinusoidal waveform (Fig. 6.1(a))

$$\begin{array}{l}\text{mean value}\\\text{of current}\end{array} = \frac{2 \times \text{maximum value of current}}{\pi} \qquad [5]$$

$$\begin{array}{l}\text{mean value}\\\text{of voltage}\end{array} = \frac{2 \times \text{maximum value of voltage}}{\pi} \qquad [6]$$

For a square waveform (Fig. 6.2(a))

$$\text{mean value of current} = \text{maximum value of current} \qquad [7]$$

$$\text{mean value of voltage} = \text{maximum value of voltage} \qquad [8]$$

A quantity which is used to give an indication of the form of a waveform is the *form factor*. This is defined as

$$\text{form factor} = \frac{\text{r.m.s. value}}{\text{mean value of half a cycle}} \qquad [9]$$

Thus for a sinusoidal waveform

$$\text{form factor} = \frac{(\text{max. value}/\surd 2)}{(2 \times \text{max. value}/\pi)} = \frac{\pi}{2\surd 2} = 1.11 \qquad [10]$$

For a square waveform

$$\text{form factor} = \frac{(\text{max. value})}{(\text{max. value})} = 1.00 \qquad [11]$$

**Moving coil meter: principles**

The *permanent magnet moving coil meter* (Fig. 6.3) consists of a coil situated in a constant magnetic field provided by a permanent magnet. When a current is passed through the coil it rotates, the angle through which it rotates being proportional to the current. This arrangement is sometimes referred to as a *D'Arsonval movement*.

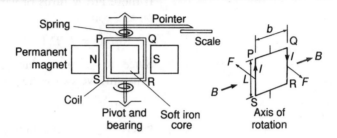

**Fig. 6.3** Moving coil meter

When a current flows through the coil, forces act on it because the sides of the coil are current-carrying conductors in a magnetic field. The force on a side of length $L$ is

$$F = BIL \qquad\qquad [12]$$

where the force $F$ is in newtons (N) when the magnetic flux density at right angles to the wire is $B$ in tesla (T) and the current $I$ is in amps (A). The force is at right angles to both the wire and the magnetic field (*Fleming's left-hand rule*: if the first finger of the left hand represents the direction of the magnetic field, the second finger the current direction, and they are held at right angles to each other and the thumb, then the direction of the thumb represents the direction of the force). It is only the forces acting on the sides QR and PS that are of interest since these are the only ones that can cause the coil to rotate. The force acting on each wire in the coil side QR or PS is $BIL$. The magnetic field $B$ is so designed that it is always at right angles to the sides QR and PS of the coil no matter what angle the coil has rotated through. The forces on the wires in side QR and PS are in opposite directions because the currents in the two sides are in opposite directions. As a consequence the forces both cause rotation of the coil about its central axis. The turning moment or torque of a force about an axis is the product of the force size and its perpendicular distance from the axis. Thus for the force acting on a wire in side QR, since for a coil of width $b$ the perpendicular distance is $\frac{1}{2}b$, the turning moment or torque is $F(\frac{1}{2}b)$ and so

$$\text{torque} = BIL(\tfrac{1}{2}b)$$

Similarly for the torque of the force acting on a wire in side PS the torque is $BIL(\frac{1}{2}b)$. Since both the sides QR and PS will contribute torques in the same direction the total turning moment or torque acting on each wire is

$$\text{torque} = BIL \times \tfrac{1}{2}b + BIL \times \tfrac{1}{2}b = BILb$$

Since $Lb$ is the area $A$ of the coil, then

$$\text{torque} = BIA$$

If there are $N$ turns of wire in the coil then the torque acting on the coil is

$$\text{torque} = NBIA \qquad\qquad [13]$$

Since for a particular galvanometer $N$, $B$ and $A$ will be constant, we can write

$$\text{torque} = K_c I$$

where $K_c$ is a constant ($NBA$) for that galvanometer. This torque is proportional to the current and causes the coil to rotate.

The rotation of the coil is against the resisting forces of springs. The springs generate an opposing torque which is proportional to the angle through which the coil rotates:

$$\text{torque due to springs} = K_s \theta$$

where $K_s$ is a constant. The coil thus rotates under the action of the torque produced by the current through it until the torque due to the springs rises to equal it. Thus when this occurs and there is no net torque acting on the coil we have $K_c I = K_s \theta$ and so

$$\theta = (K_c/K_s)I \qquad\qquad [14]$$

The angular deflection $\theta$ of the coil is thus proportional to the current $I$.

The magnetic field is produced by a permanent magnet, made of a material such as Alcomax, Alnico or Columax. The pole pieces are so shaped that with a central cylindrical soft iron core a magnetic field is produced which is always at right angles to the sides of the coil. The coil in the typical instrument is wound on a copper or aluminium former mounted on jewel bearings so that it is free to rotate. When this former moves in the magnetic field eddy currents are induced in it. But forces act on current-carrying conductors in magnetic fields. The direction of the forces acting on these currents as a consequence of being in the magnetic field is such as to oppose the motion producing them. As a result the presence of the eddy currents slows down the motion of the coil and is said to provide damping. The springs used to provide the torque opposing the rotation of the coil are generally flat or helical phosphor–bronze springs. The springs also provide the means by which the current is supplied to the coil.

## Example 1

Which of the following when increased will increase the sensitivity of a moving coil galvanometer, i.e., increase the angular deflection for a given current?
(a) The magnetic flux density provided by the permanent magnet.
(b) The cross-sectional area of the coil.
(c) The torque constant $K_s$ of the springs.

*Answer*

The sensitivity is $\theta/I$ and thus using the equations developed above

$$\frac{\theta}{I} = \frac{K_c}{K_s} = \frac{NBA}{K_s}$$

(a) Increasing the flux density $B$ increases the sensitivity.
(b) Increasing the cross-sectional area $A$ increases the sensitivity.
(c) Increasing the torque constant $K_s$ decreases the sensitivity.

**Moving coil meter: current range extension**

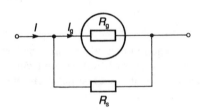

**Fig. 6.4**   Instrument shunt

The current required to give a full-scale deflection with the basic permanent magnet moving coil meter is generally in the range of $10\,\mu A$ to $20\,mA$. *Shunts* are resistors connected in parallel with a meter in order to change the current range (Fig. 6.4). When components are in parallel then the potential difference across each component is the same and the total current entering the arrangement equals the sum of the currents through each component. If $I_g$ is the full-scale current possible for the basic meter then, using Ohm's law ($V = IR$), the potential difference across the meter is $I_g R_g$, where $R_g$ is the resistance of the meter. If $I$ is the total current entering the arrangement, i.e., the current to be measured, then since $I$ is the sum of the current through the meter and that through the shunt, the current through the shunt is

current through shunt $= I - I_g$

Thus, using Ohm's law ($V = IR$), the potential difference across the shunt is

potential difference across shunt $= (I - I_g)R_s$

Thus, since the potential difference across the shunt equals the potential difference across the meter

$$(I - I_g)R_s = I_g R_g$$

$$I = \frac{I_g(R_s + R_g)}{R_s} \qquad [15]$$

Thus $(R_g + R_s)/R_s$ is the scaling factor by which the range of the meter is changed.

While a multi-range instrument can be made by switching

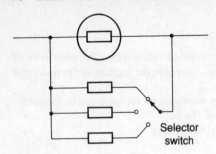

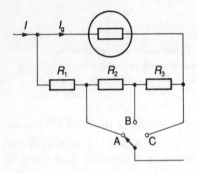

**Fig. 6.5**  Multi-range shunt

**Fig. 6.6**  Universal shunt

into circuit a shunt resistor from a selection of resistors, as in Fig. 6.5, another possibility is the *universal shunt* (Fig. 6.6), sometimes referred to as the *Ayrton shunt*. With the switch in the position A shown in Fig. 6.6 resistor $R_1$ is shunting the meter, while $R_2$ and $R_3$ are in series with it and so added to the meter resistance. The potential difference across the shunt $R_1$ must be the same as the potential difference across the meter plus its series resistors, i.e., a total resistance of $(R_g + R_2 + R_3)$, and since the current through the meter and its resistors is $I_g$ and that through the shunt $(I - I_g)$ then using Ohm's law $(V = IR)$

$$(I - I_g)R_1 = I_g(R_g + R_2 + R_3) \qquad [16]$$

With the switch in position B, resistors $R_1$ and $R_2$ are shunting the meter and $R_3$ in series with it. In this case we then have

$$(I - I_g)(R_1 + R_2) = I_g(R_g + R_3) \qquad [17]$$

With the switch in position C, all the resistors shunt the meter. Thus

$$(I - I_g)(R_1 + R_2 + R_3) = I_g R_g \qquad [18]$$

**Example 2**

What is the resistance of the shunt required to convert a moving coil instrument which gives a full-scale deflection of 10 mA to one giving a full-scale deflection of 10 A? The instrument has a resistance of 40 Ω.

*Answer*

Using the equation [15] and rearranging it,

$$R_s = \frac{I_g R_g}{I - I_g} = \frac{0.010 \times 40}{10 - 0.010} = 0.040\,\Omega$$

**Example 3**

Calculate the values of the resistors to be used in a universal shunt for a three-range meter. The basic meter has a full-scale deflection of 1.0 mA and a resistance of 50 Ω, and the required ranges are 10 mA, 100 mA and 1.0 A.

*Answer*

The circuit for the universal shunt is as shown in Fig. 6.6. The maximum range will be when just one of the resistors is shunting the meter and the minimum range when all three are. This is because the maximum range occurs when the smallest resistance shunts the meter and so the smallest fraction of the current passes through the meter. For just $R_1$ shunting the meter and $R_2$ and $R_3$ in series with it (switch in position A in Fig. 6.6) using equation [16]

$$(I - I_g)R_1 = I_g(R_g + R_2 + R_3)$$

$$(1.0 - 0.001)R_1 = 0.001(50 + R_2 + R_3)$$

$$999R_1 = 50 + R_2 + R_3 \qquad [19]$$

For $R_1$ and $R_2$ shunting the meter and $R_3$ in series with it (switch in position B in Fig. 6.6), equation [17] gives

$$(I - I_g)(R_1 + R_2) = I_g(R_g + R_3)$$

$$(0.100 - 0.001)(R_1 + R_2) = 0.001(50 + R_3)$$

$$99(R_1 + R_2) = 50 + R_3 \qquad [20]$$

For $R_1$, $R_2$ and $R_3$ shunting the meter and no resistors in series with it (switch in position C in Fig. 6.6), equation [18] gives

$$(I - I_g)(R_1 + R_2 + R_3) = I_gR_g$$

$$(0.010 - 0.001)(R_1 + R_2 + R_3) = 0.001 \times 50$$

$$9(R_1 + R_2 + R_3) = 50 \qquad [21]$$

We now have three simultaneous equations [19], [20] and [21]. If we multiply equation [21] by 11 we obtain

$$99R_1 + 99R_2 + 99R_3 = 550$$

and so

$$99R_1 + 99R_2 = 550 - 99R_3$$

Thus, using equation [20] we have

$$550 - 99R_3 = 50 + R_3$$

Hence $R_3 = 5.0\,\Omega$. Multiplying equation [21] by 111 gives

$$999R_1 + 999R_2 + 999R_3 = 5550$$

and so

$$999R_1 = 5550 - 999R_2 - 999R_3$$

Thus this with equation [19] gives

$$5550 - 999R_2 - 999R_3 = 50 + R_2 + R_3$$

Since $R_3 = 5.0\,\Omega$ then

$$5550 - 999R_2 - 4995 = 50 + R_2 + 5$$

Hence $R_2 = 0.50\,\Omega$. We can obtain the value of $R_1$ by substituting these values of $R_2$ and $R_3$ in equation [21],

$$9R_1 + 9 \times 0.50 + 9 \times 5 = 50$$

Hence $R_1 = 0.056\,\Omega$.

**Moving coil meter: voltage range extension**

The basic moving coil meter in responding to the current through the meter coil is responding to the current through a fixed resistance, that of the coil. Since $V = IR$, the response of the meter is also proportional to the potential difference across the meter and so is a voltmeter. However, because the meter resistance is relatively low and the current it responds to is

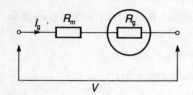

**Fig. 6.7** Instrument multiplier

low, the basic meter can only be used for low voltages. Thus for a typical meter the resistance might be $50\,\Omega$ and the current to give full-scale deflection $1\,mA$, hence the voltage to give full-scale deflection would be $1 \times 10^{-3} \times 50 = 0.05\,V$.

Resistors connected in series with an instrument can be used to change the voltage range of that instrument, such resistors being called *multipliers* (Fig. 6.7). If $V$ is the full-scale voltage which is required to be measured then $V$ has to be the sum of the potential differences (p.d.) across both the multiplier and the instrument.

$$V = \text{p.d. across multiplier} + \text{p.d. across meter}$$

The full-scale voltage will occur for the full-scale current $I_g$ through the meter. Because the multiplier and meter are in series then the current through the multiplier will also be $I_g$. Hence, since the potential difference across the multiplier is $I_g R_m$, with $R_m$ being the resistance of the multiplier, and the potential difference across the meter is $I_g R_g$, with $R_g$ being the resistance of the meter, then

$$V = I_g R_m + I_g R_g$$

and so

$$V = I_g(R_m + R_g) \qquad [22]$$

Multi-range voltmeters can be produced by switching different resistors in series with the meter (Fig. 6.8(*a*)). An alternative to this switching of separate resistors in series with the meter is to use a chain arrangement by which additional resistors are switched in series with the meter (Fig. 6.8(*b*)).

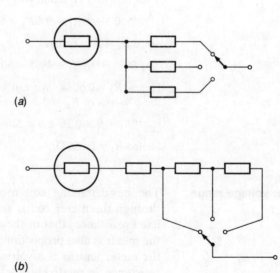

(a)

(b)

**Fig. 6.8** Multi-range multiplier: (a) separate, (b) chain

**Example 4**

A moving coil meter gives a full-scale deflection of 1.0 mA and has a resistance of 80 Ω. What is the resistance of the multiplier required to convert this instrument to a meter giving a full-scale deflection with 10 V?

*Answer*

Using equation [22]

$$V = I_g(R_m + R_g) = 10 = 1.0 \times 10^{-3}(R_m + 80)$$

Hence $R_m = 9920\,\Omega$.

## Moving coil meter: effect of temperature

The meter coil is wound with copper wire and thus when there is a change in temperature, its resistance changes. A rise in temperature produces an increase in resistance and so causes the meter to read low. This is partly compensated for by a decrease in spring tension which occurs with an increase in temperature. However, the meter is likely to read low by about 0.2% per °C rise in temperature. To reduce this a resistor is often connected in series with the coil, this resistor being known as a *swamp resistor*. This resistor is usually made of manganin wire, a material whose resistance change with temperature is relatively low, and has a resistance about three times that of the coil. The change in resistance produced by a temperature change thus becomes a smaller percentage of the total resistance and so the effect of the temperature change on the current through the meter is reduced.

Where there is a shunt the arrangement used is as shown in Fig. 6.9. With a shunt made of copper wire its resistance will increase with an increase in temperature and since that of the manganin series swamp resistor does not change, a greater proportion of the total current flows through the meter. This compensates for the reduction that otherwise would have occurred in the absence of the swamp resistor. The disadvantage of using swamp resistors is that they result in a reduction in the sensitivity of the meter since the total resistance has been increased.

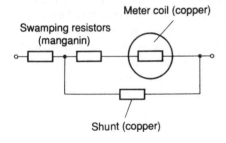

Meter coil (copper)

Swamping resistors (manganin)

Shunt (copper)

**Fig. 6.9** Swamp resistor

## Moving coil meter: sensitivity

In general the sensitivity of an instrument can be defined as

$$\text{sensitivity} = \frac{\text{change in position of pointer}}{\text{change in input to the instrument}} \qquad [23]$$

Thus the *sensitivity* S of a meter can be defined as the angular deflection of the pointer per unit current input, i.e.,

$$S = \frac{\theta}{I}$$

However, because the angular deflection needed to give a full-scale deflection is much the same for many meters, the sensitivity is often just defined as

$$S = \frac{1}{I_{fsd}}$$

where $I_{fsd}$ is the current to give full-scale deflection. There is an alternative form of expressing this definition which is the customary way of expressing the sensitivity for moving coil meters. If the meter has a resistance $R$ then the full-scale deflection potential difference across it of $V_{fsd}$ is $I_{fsd}R$, hence

$$S = \frac{R}{V_{fsd}} \qquad [24]$$

For this reason the unit often used for sensitivity is $\Omega/V$.

The ohm/volt rating can be considered a measure of the size of error that will occur as a result of loading (see Chapter 3) when a moving coil voltmeter is used to measure a voltage. The percentage error introduced when a meter with resistance $R_m$ is used with a circuit having a Thévenin equivalent resistance of $R_{Th}$ is (equation [5], Chapter 3)

$$\text{percentage error} = -\frac{R_{Th}}{R_m + R_{Th}} \times 100\%$$

Thus the higher the meter resistance, the smaller the error. An ideal voltmeter would have an infinite resistance and hence an infinite ohm/volt value. Such a voltmeter when connected to a circuit would not affect the circuit resistance and so would produce no loading error. The smaller the ohm/volt value the smaller will be the voltmeter resistance for a given measurement and so the greater will be the loading error.

### Example 5

A moving coil meter has a resistance of $40\,\Omega$. When used as a voltmeter with a multiplier of resistance $200\,k\Omega$, a full-scale deflection is obtained with $300\,V$. What is the sensitivity of the voltmeter?

*Answer*

Using equation [24]

$$S = \frac{R}{V_{fsd}}$$

where $R$ is the total resistance of the meter and its multiplier. Thus

$$S = \frac{200\,040}{300} = 666.8\,\Omega/V$$

**Example 6**

What is the percentage error that would be produced in using a voltmeter with a sensitivity of $1000\,\Omega/V$ to measure, using its $50\,V$ scale, the voltage in a circuit having a Thévenin resistance of $10\,k\Omega$?

*Answer*

The resistance of the meter on its $50\,V$ scale is $50\,k\Omega$. Thus, using the equation given above,

$$\text{percentage error} = -\frac{R_{Th}}{R_m + R_{Th}} \times 100\%$$

$$= -\frac{10}{50 + 10} \times 100\% = -17\%$$

**Moving coil meter: a.c. measurement**

Because the torque acting on the coil of a moving coil meter depends on the size and direction of the applied current at the instant concerned, an alternating current will lead to an alternating torque. For all but very low frequencies the mechanical inertia of the conventional meter coil will be such that the current will be changing too fast for the coil to keep up with it and so it ends up not moving from the zero position. To enable a moving coil meter to give a measure of an alternating current it is necessary to convert it to a direct current. Figure 6.10 shows a basic circuit that can be used. The rectifiers are usually germanium or silicon diodes. The bridge circuit converts an alternating current input into a pulsating unidirectional current through the meter. Because of the inertia of the coil of the meter it indicates the average value of the pulses. These values are usually converted to the equivalent root-mean-square values, and the scale marked in these values, on the assumption that the input to the bridge is a sinusoidal alternating current.

For a sinusoidal signal the root-mean-square current $I_{rms} = I_m/\sqrt{2}$, where $I_m$ is the maximum or peak current value. The

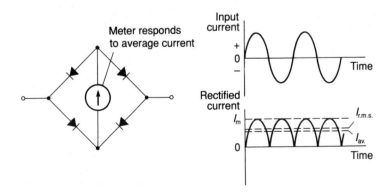

**Fig. 6.10** Moving coil meter for a.c.

average current $I_{av} = 2I_m/\pi$. Thus

$$I_{rms} = \frac{\pi I_{av}}{2\sqrt{2}} = 1.11 I_{av}$$

The r.m.s. value $I_{rms}$ is thus 1.11 times the average value $I_{av}$ and the instrument scale is calibrated on this basis. If the input is not sinusoidal then the scale is in error by the amount the ratio $I_{rms}/I_{av}$ differs from 1.11. This ratio is called the form factor $F$. The error is thus

$$error = \left(\frac{1.11 - F}{F}\right) \times 100\% \qquad [25]$$

Different current ranges can be obtained by using a current transformer with different tappings for the various ranges. The moving coil meter adapted for a.c., as in Fig. 6.10, can be used for the measurement of alternating voltages by including a resistor in series with the bridge, i.e., a multiplier. As with its use for alternating current, the meter is calibrated in terms of the r.m.s. voltage value, on the assumption that the input is sinusoidal. The reading will be in error if this is not the case and the correction is as indicated above.

### Example 7

What is the error that will occur if a moving coil meter is used to measure a square wave alternating current if the meter is calibrated assuming a sinusoidal alternating current? A square wave has a form factor of 1.00.

*Answer*

Using equation [25]

$$error = \left(\frac{1.11 - F}{F}\right) \times 100\% = \frac{(1.11 - 1.00)}{1.00} \times 100\% = +11\%$$

This means that the reading indicated is 11% high.

### Example 8

A moving coil meter has a resistance of $50\,\Omega$ and gives a full-scale deflection with $1\,mA$. The meter is to be converted, using a bridge rectifier circuit, to give a full-scale deflection with an alternating voltage of $10\,V$ (r.m.s.). If the rectifiers can be assumed to have zero forward resistance and infinite reverse resistance, what is the resistance of the multiplier that has to be connected in series with the bridge rectifier?

*Answer*

The moving coil meter with the bridge rectifier responds to the average value of the rectified pulses. The average value $V_{av}$ is $2V_m/\pi$ and the root-mean-square value $V_{rms}$ is $V_m/\sqrt{2}$. Hence

$$V_{av} = \frac{2\sqrt{2}}{\pi} V_{rms} = 0.90 V_{rms}$$

With $V_{rms} = 10\,V$, the average voltage is $9.0\,V$. This is to be the potential difference across the meter, the bridge rectifier and the multiplier. Since the resistance of the bridge rectifier can be assumed, from the information given, to be zero then

$$V_{av} = I_{fsd}(R_g + R_m)$$

where $I_{fsd}$ is the full-scale current, $R_g$ the resistance of the meter and $R_m$ the resistance of the multiplier. Hence

$$9.0 = 1 \times 10^{-3}(50 + R_m)$$

$$R_m = 8950\,\Omega$$

**Moving coil meter: performance**

The accuracy of a moving coil meter is generally about $\pm 0.1$–$5\%$ and is affected by such factors as temperature, the way the meter is mounted, friction at the bearings, parallax errors and errors from interpolating between scale marking in reading the meter. The time taken for a steady deflection to be obtained is typically in the region of a few seconds. The moving coil meter has a linear scale and basically gives a full-scale deflection with a direct current of between $10\,\mu A$ and $20\,mA$. The measurement of larger currents, up to about $10\,A$, involve the use of shunts. For alternating currents, full-scale deflections vary from about $10\,mA$ to $10\,A$. As a voltmeter, full-scale deflections for direct voltages from about $50\,mV$ to $100\,V$ can be obtained with multipliers. For alternating voltages, full-scale deflections vary from about $3\,V$ to $3000\,V$. The frequency range within which alternating current or voltage measurements can be made is about $50\,Hz$ to $10\,kHz$. The instrument has a linear scale but is only calibrated to give correct r.m.s. values for sinusoidal alternating currents and voltages, corrections have to be made for other forms (see earlier in this chapter). The sensitivity is typically about $20\,000\,\Omega/V$ for all d.c. ranges and for a.c. varies from $100\,\Omega/V$ for voltage ranges of about 0–3 V to $2000\,\Omega/V$ for ranges of more than 0–10 V. Loading effects (see Chapter 3) can present problems.

**Example 9**

A voltmeter is to be used to measure the potential difference across a circuit with an effective resistance of $1\,M\Omega$. What should be the resistance of the voltmeter if the loading error is to be kept to less than $\times 5\%$?

*Answer*

Using equation [5] from Chapter 3

$$\text{percentage error} = -\frac{R_{\text{Th}}}{R_{\text{m}} + R_{\text{Th}}} \times 100\%$$

where $R_{\text{Th}}$ is the Thévenin equivalent resistance of the circuit and $R_{\text{m}}$ the resistance of the meter. Thus

$$\frac{95}{100} = -\frac{10^6}{R_{\text{m}} + 10^6}$$

Hence $R_{\text{m}} = 52.6 \, \text{k}\Omega$

**Moving coil meter: ohmmeter**

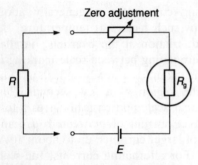

Zero adjustment

**Fig. 6.11** Series type ohmmeter

The term *ohmmeter* is used for an arrangement involving a moving coil meter which can be used for the measurement of resistance. Figure 6.11 shows one form of a basic ohmmeter circuit, a so-called *series type ohmmeter*. A battery, e.m.f. $E$, is connected in series with the meter, total resistance $R_{\text{g}}$, and the resistance $R$ to be measured. The current $I$ through the meter depends on the resistance in the circuit and thus can be used to obtain a measurement of the resistance $R$. A variable resistor is generally included in the circuit to compensate for changes in the e.m.f. of the battery. The procedure for using this instrument is:

1   With the terminals of the instrument short-circuited, i.e., $R = 0$, the 'zero adjustment' resistor is adjusted so that there is a full-scale current reading, i.e., with a meter having a resistance scale this means that the resistance reading is zero.

2   The unknown resistance $R$ is then connected across the meter terminals. The current through the meter then diminishes because the total circuit resistance has increased. The value of the current is a measure of the resistance of $R$ and thus the pointer position on the scale can be used to give the resistance value.

For the series type ohmmeter, because the resistors are in series

$$E = I(R + R_{\text{g}} + R_{\text{z}})$$

where $R_{\text{z}}$ is the resistance of the 'zero adjustment' resistor. Thus

$$I = \frac{E}{R + R_{\text{g}} + R_{\text{z}}} \qquad [26]$$

The relationship between the current and the resistance $R$ is non-linear and hence the ohmmeter circuit gives a non-linear resistance scale with the zero of the resistance scale corresponding to a full-scale current reading. The scale points are very close together for low resistances and so this form of ohmmeter is not very useful for measuring them.

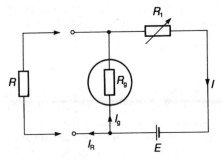

**Fig. 6.12** Shunt type ohmmeter

Figure 6.12 shows another form of basic ohmmeter circuit, a so-called *shunt type ohmmeter*. The procedure for using this meter is:

1  When $R$ is infinity, i.e., open-circuit, resistance $R_1$ is adjusted to give a full-scale reading on the meter.
2  The resistance $R$ is connected into the circuit. This acts as a shunt for the meter and so reduces the current passing through the meter. The meter reading is thus a measure of the resistance $R$.

When $R$ is open circuit and $R_1$ adjusted to give the full-scale deflection current $I_{fsd}$, then

$$E = I_{fsd}(R_1 + R_g)$$

and so

$$I_{fsd} = \frac{E}{R_1 + R_g}$$

The unknown resistance is then connected in parallel with the meter and acts as a shunt, some of the current that would have passed through the meter now passing through the resistor. Thus the resistance $R_{sm}$ of the meter in parallel with $R$ is given by

$$\frac{1}{R_{sm}} = \frac{1}{R} + \frac{1}{R_g}$$

$$R_{sm} = \frac{RR_g}{R + R_g}$$

The current $I$ in the circuit is thus given by

$$E = I(R_1 + R_{sm})$$

$$E = I\left(R_1 + \frac{RR_g}{R_g + R}\right)$$

The current $I$ divides into two parts, part passing through the meter and part through the resistor $R$. Thus the current through the meter $I_g$ is

$$I_g = I - I_R$$

where $I_R$ is the current through $R$. But $I_g R_g = I_R R$ and so

$$I_g = I - \frac{R_g I_g}{R}$$

Hence

$$I_g = \frac{ER}{R_1 R_g + R(R_1 + R_g)} \tag{27}$$

The meter reading thus depends on the value of the unknown resistance $R$. This form of ohmmeter is particularly useful for low resistances.

### Example 10

A series type ohmmeter uses a moving coil meter with a resistance of $50\,\Omega$ and a full-scale deflection of $1.0\,\text{mA}$. If the ohmmeter battery has an e.m.f. of $2.0\,\text{V}$ what should be the value of the resistor in series for the meter for the zero adjustment and the shunt required for the meter if the mid-scale reading of the ohmmeter is to be $1000\,\Omega$?

*Answer*

When the terminals of the ohmmeter are short-circuited to give the zero resistance reading, i.e., full-scale current, then

$$2.0 = I(R_g + R_z)$$

where $I$ is the full-scale current through the shunted meter, $R_g$ is the resistance of the shunted meter and $R_z$ is the resistance of the series resistor adjusted for the zero resistance reading. When there is a resistance of $1000\,\Omega$ between the ohmmeter terminals the current is $\frac{1}{2}I$. Hence

$$2.0 = \tfrac{1}{2}I(R_g + R_z + 1000)$$

$$4.0 = I(R_g + R_z) + 1000I = 2.0 + 1000I$$

Hence the full-scale current $I$ through the shunted meter is $2.0\,\text{mA}$. Thus since the meter gives a full-scale deflection with $1.0\,\text{mA}$, the current through the shunt must be $1.0\,\text{mA}$. This means the shunt must have the same resistance as the meter, i.e., $50\,\Omega$. Consequently $R_z$ is $950\,\Omega$.

## Moving coil multimeter

The basic moving coil meter movement can be used with appropriate shunts for the measurement of a number of ranges of direct currents, with appropriate multipliers for a number of ranges of direct voltages and with a rectifier circuit for alternating currents and voltages. The meter can also be used as the basis of an ohmmeter. All these functions can be combined to give a moving coil multimeter, i.e., a multi-range instrument which can be used for the measurement of direct and alternating currents and voltages, and resistance. A typical meter has full-scale deflections for d.c. current ranges from $50\,\mu\text{A}$ to $10\,\text{A}$, a.c. current $10\,\text{mA}$ to $10\,\text{A}$, d.c. voltage $100\,\text{mV}$ to $3000\,\text{V}$, a.c. voltage $3\,\text{V}$ to $3000\,\text{V}$, resistance $2\,\text{k}\Omega$ to $20\,\text{M}\Omega$, with d.c. accuracy of $\pm1\%$ f.s.d., a.c. accuracy $\pm2\%$ f.s.d. and resistance $\pm3\%$ of the mid scale reading.

## Electronic multimeters

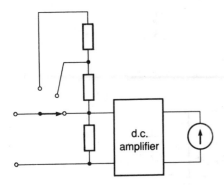

**Fig. 6.13**  Direct voltage range circuit

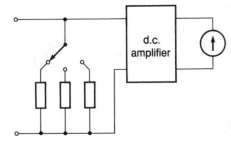

**Fig. 6.14**  Direct current range circuit

The moving coil meter has the disadvantages of a low input impedance, hence the greater likelihood of loading errors, and low sensitivity when used with a rectifier circuit for a.c. ranges. These limitations can be overcome by using an amplifier between the input and the moving coil meter, such an arrangement being referred to as an electronic meter.

Figure 6.13 shows the basic voltage range circuit for the measurement of direct voltages. A chain of resistors is used with additional resistors being switched in series with the amplifier as multipliers. Such a meter would typically have an input impedance of about $10\,\text{M}\Omega$, an accuracy of about $\pm 1\%$ and full-scale deflections ranging from about $15\,\text{mV}$ to $1000\,\text{V}$. For direct current measurement a series of switched shunts is generally used (Fig. 6.14). Typically, ranges extend from full-scale deflections of $1\,\mu\text{A}$ to $3\,\text{A}$ with an accuracy of about $\pm 1$–$2\%$ of full-scale reading and an input resistance which decreases from a few thousand ohms on the low current range to less than an ohm on the high current range.

There are a number of versions of electronic voltmeter for alternating voltage determination. The *mean* or *average-responding type* gives an output which is a measure of the average value of the rectified voltage waveform. With a d.c. amplifier this type has a bridge rectifier to convert the alternating voltage into a series of pulses before the amplification (Fig. 6.15). The scale is usually calibrated in r.m.s. values, assuming the input signal to be sinusoidal. If this is not the case the scale is in error and corrections have to be made according to the form factor of the input (see earlier in this chapter). The *peak-responding type* gives an output which is a measure of the peak value of the voltage waveform. The input is half-wave rectified and then applied to a capacitor to charge it to the peak value of the voltage (Fig. 6.16). This peak value is then amplified. The *r.m.s.-responding type* gives an output which is a measure of the r.m.s. value of the voltage waveform, regardless of its waveform. One form of such an instrument has a thermocouple which gives an output related to the temperature of a resistor. Since the heat developed in the resistor is proportional to the power dissipated in the resistor ($V^2/R$), the thermocouple output is a measure of the

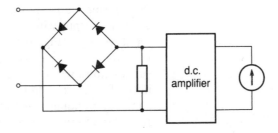

**Fig. 6.15**  Average-responding electronic voltmeter

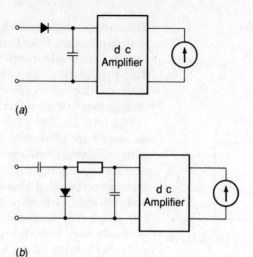

(a)

**Fig. 6.16** Peak-responding electronic voltmeter

(b)

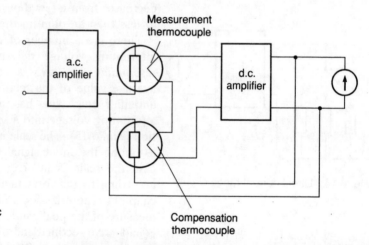

Measurement thermocouple

Compensation thermocouple

**Fig. 6.17** RMS-responding electronic voltmeter

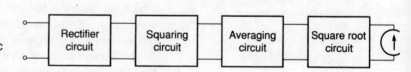

**Fig. 6.18** RMS-responding electronic voltmeter

average value of the square of the voltage. The output from the thermocouple is then amplified by a d.c. amplifier and gives an output which is a measure of the r.m.s. value of the input voltage (Fig. 6.17). The effect of non-linearity in the response of the measuring thermocouple can be cancelled by similar non-linearity in the response of another thermocouple in the feedback circuit of the amplifier. An alternative form of r.m.s.-responding voltmeter (Fig. 6.18) involves circuits to rectify the input voltage, then to square the voltage, average it over a cycle and then take the square root and feed it to a moving coil meter. Electronic voltmeters typically have full-scale deflections from about 100 μV to 1000 V, input impedance of about 10 MΩ, can be used with frequencies from about 20 Hz to 100 MHz, and an accuracy which varies from about ±2–5% depending on the range and the frequency.

Alternating current is measured by using an electronic voltmeter to determine the voltage drop across a known resistor. Typically ranges extend from full-scale deflections of 1 μA to 3 A with an accuracy of ±2–5%, this depending on the range and the frequency, measurements being possible from about 20 Hz to 100 MHz.

Unknown resistances are determined by measuring the voltage drop across them when supplied with a constant current (Fig. 6.19). Different resistance ranges can be obtained by using the different ranges of the direct voltage electronic voltmeter. Typically such an arrangement is used to measure resistances, in a series of ranges, from 1 Ω to 100 MΩ. The scales are non-linear and have an accuracy of about ±3% of the mid-scale value.

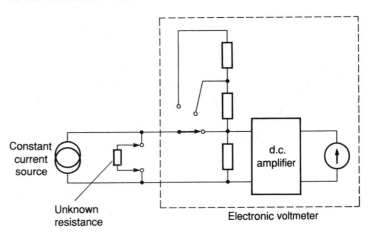

**Fig. 6.19**   Resistance measurement with an electronic voltmeter

**Light spot galvanometer**

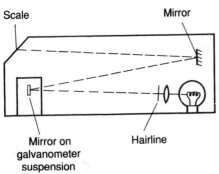

**Fig. 6.20**   Light spot galvanometer

For the measurement and detection of very small currents, e.g., as a zero current detector in a Wheatstone bridge (see Chapter 9), a *light spot galvanometer* can be used (Fig. 6.20). This is a form of moving coil meter, the coil being suspended by a metal strip or wire, the twisting of this providing the restoring torque rather than springs. Rotation of the coil causes a beam of light to be reflected across a scale. Typically, such instruments have current sensitivities of the order of 100 mm of movement across the scale per microamp.

**Moving iron meter: principles**

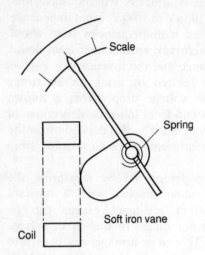

**Fig. 6.21** Attraction type moving iron meter

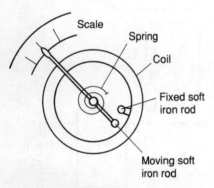

**Fig. 6.22** Repulsion type moving iron meter

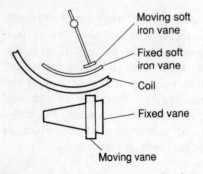

**Fig. 6.23** Repulsion type moving iron meter

There are two basic types of moving iron meter, one based on magnetic attraction and the other on magnetic repulsion. In both cases the current to be measured is passed through a coil of wire. This results in the production of a magnetic field, the strength of the field being proportional to the current $I$ through the coil. With the *attraction type* of instrument the magnetic field of the coil attracts a pivoted soft iron disk towards it (Fig. 6.21). The torque resulting from this attraction is proportional to the square of the current through the coil, i.e.,

$$\text{torque} = kI^2$$

where $k$ is a constant. This torque is opposed by a torque resulting from springs, the torque being proportional to the angle $\theta$ through which the disk and pointer have rotated,

$$\text{restoring torque} = k_s\theta$$

where $k_s$ is a constant for the springs. At equilibrium $k_s\theta = kI^2$ and thus

$$\theta = (k/k_s)I^2 \qquad [28]$$

Thus the angular deflection is proportional to the square of the current.

The *repulsion type* has two pieces of soft iron inside the current-carrying coil. One of the pieces is fixed and the other able to move, being fixed to the end of a pivoted pointer. Figure 6.22 shows one form of such a meter where the pieces of soft iron are rods, and Fig. 6.23 shows another form where the soft iron is in the form of two vanes. When a current passes through the coil both pieces of iron become magnetized in the same way and so repulsion occurs. The repulsion force depends on the extent to which the two pieces of soft iron have become magnetized and this in turn depends on the magnetic field produced by the current through the coil and hence the current. The deflecting torque is proportional to the square of the current $I$, i.e.

$$\text{torque} = kI^2$$

where $k$ is a constant. A restoring torque is provided by springs, this being proportional to the angle $\theta$ through which they are twisted. Thus

$$\text{restoring torque} = k_s\theta$$

where $k_s$ is a constant for the springs. At equilibrium $k_s\theta = kI^2$ and thus

$$\theta = (k/k_s)I^2 \qquad [29]$$

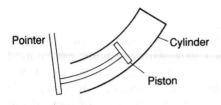

Fig. 6.24 Air damping

The result is that the angular deflection is proportional to the square of the current.

Because the deflection of a moving iron meter is proportional to the square of the current, the instrument has a non-linear scale, the readings being cramped at the lower end. However, a consequence of the deflection being proportional to the square of the current is that the meter can be used with both direct and alternating current. With alternating current, the meter gives a steady deflection proportional to the mean of the squares of the current; the scale is thus usually calibrated in terms of the root-mean-square current. Damping in both types of moving iron meter is usually provided by an air damper involving a piston moving in a cylinder (Fig. 6.24).

### Example 11

A moving iron meter gives a full-scale deflection with 1.0 A. What will be the current for the mid-scale reading?

*Answer*

Since the deflection is proportional to the square of the current, the mid-scale reading will be given by

$$\frac{\theta_{mid}}{\theta_{fsd}} = \frac{I^2_{mid}}{I^2_{fsd}}$$

$$I_{mid} = \frac{I_{fsd}}{\sqrt{2}} = 0.71 \, \text{A}$$

**Moving iron meter: range extension**  Shunts can be used to extend the range of a moving iron meter when it is used for direct currents but not for alternating currents. When used with a.c., resistive shunts can lead to errors as a result of changes in the coil reactance with frequency, the coil being effectively an inductance in series with a resistance. This effect can be reduced by using a 'swamp' resistor in series with the instrument, or a capacitor shunt, or a shunt with an inductance to resistance ratio equal to that of the coil. A more usual way, however, is for the meter coil to be made up of a number of coils which can be connected in various series–parallel combinations to give the different current ranges. Since the strength of the magnetic field produced by a current through a coil depends on the product of the number of turns and the current, the smaller the current, the greater the number of turns required to obtain the same magnetic field and give a full-scale deflection. Thus for the measurement of small currents the meter needs a large number of turns and as a consequence has a relatively high resistance.

The range of the moving iron meter when used as a

voltmeter can be extended by connecting non-inductive resistances in series and, when used with alternating voltages, a voltage transformer.

### Moving iron meter: performance

Typically, moving iron meters have ranges with full-scale deflections between 0.1 and 30 A, without shunts, can be used for both d.c. and a.c., and have an accuracy of about ±0.5% of full-scale deflection. Errors are produced by changes in temperature, since these change the resistance of the coil and the permeability of the iron. When used with d.c. the instrument gives a low indication of a slowly increasing current and a high indication of a slowly decreasing current, this being because of the magnetic hysteresis curve of the iron used for the rods or vanes. The instrument can be used with a.c. up to a frequency of about 100 Hz; errors can, however, occur if the waveform is non-sinusoidal.

As a voltmeter the moving iron meter has a relatively low input impedance, of the order of 50 Ω/V, and a minimum full-scale deflection of about 50 V. It can be used for both direct and alternating voltages.

### Hot wire meter

This is a very simple, low accuracy, form of meter. When a current is passed through a wire it increases in temperature and expands. This expansion can be used to rotate a pointer and so give a measure of the current. The instrument is robust, cheap, can be used for both d.c. and a.c. and has a non-linear scale and low sensitivity. Errors arise due to changes in the temperature of the surroundings. The instrument is used for low accuracy applications where a cheap, robust meter is required, e.g., in cars.

### Thermocouple meter

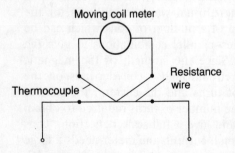

**Fig. 6.25**  Thermocouple meter

When a current is passed through a resistance wire power is dissipated and the temperature of the resistor increases. The thermocouple meter uses a thermocouple to give a measure of the temperature of a resistor (Fig. 6.25). The output from the thermocouple can then be monitored by a d.c. moving coil meter. Since the heating effect of the current is proportional to the square of the current through the resistor or the potential difference across it, the instrument can be used for alternating currents or voltages. The instrument measures the true r.m.s. values, regardless of waveshape, and can be used from about 10 Hz to as high as 50 MHz. It finds its main use in the measurement of very high frequency currents. The scale is non-linear with full-scale deflections between about 2 and 50 mA, being used with a series resistor for voltages. The instrument is fragile with a low overload capacity.

## Electrostatic meters

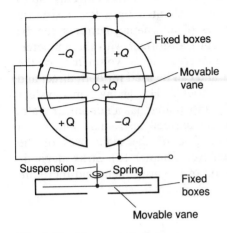

**Fig. 6.26** Electrostatic voltmeter

The basic form of the electrostatic voltmeter is of a set of four quadrant-shaped boxes in which a movable vane can rotate (Fig. 6.26). The quadrants and vane can be connected in a number of ways to the circuit for which the voltage is required. With the most commonly used form of connection, called idiostatic, opposite boxes are connected together, with the movable vane connected to one pair. The arrangement is essentially a form of capacitor. Thus when a voltage is applied, the boxes and the vane become charged. Since like charged objects repel each other and unlike ones attract, the movable vane experiences a torque. This torque is proportional to the square of the voltage. A restoring torque proportional to the angular movement of the vane is provided by springs. Thus at equilibrium the angular deflection of the vane is proportional to the square of the voltage. Consequently the instrument can thus be used for direct or alternating voltages, giving r.m.s. values regardless of the shape of the applied waveform. Electrostatic voltmeters have a high input impedance, are fragile, accurate and expensive. The range tends to be from about 100 V to 1000 V or higher. There main applications are for situations where a high voltage is to be measured and a high input impedance is vital, and as a transfer standard between a.c. and d.c. voltages.

## Electrodynamometer

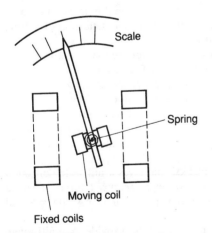

**Fig. 6.27** Electrodynamometer

The electrodynamometer is rather like a moving coil meter (see earlier in this chapter) with two fixed coils being used instead of a permanent magnet to provide the magnetic field (Fig. 6.27). It consists of a rotatable coil situated in a magnetic field provided by a pair of fixed coils. When a current is passed through the rotatable coil it rotates. The magnetic flux density $B$ produced by a current $I_1$ through the two fixed coils, is proportional to $I_1$. Situated in this magnetic field is the moving coil. The torque acting on the moving coil when it carries a current $I_2$ is proportional to $BI_2$. Hence the torque is proportional to $I_1I_2$. This deflecting torque results in the coil rotating against the restoring torque provided by springs. Since the restoring torque is proportional to the angle $\theta$ through which the coil rotates then at equilibrium

$$\theta = KI_1I_2$$

where $K$ is some constant.

With the fixed and moving coils connected in series (Fig. 6.28(a)), the current through the fixed coils is the same as that through the movable coil, i.e., $I_1 = I_2$. Then the angle through which the moving coil rotates is proportional to the square of the current. The instrument can be used for the measurement of both d.c. and a.c., giving r.m.s. values irrespective of

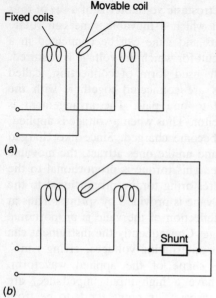

Fixed coils

Movable coil

(a)

Shunt

(b)

**Fig. 6.28** Current measurement:
(a) basic circuit, (b) with a shunt

waveform. It is used for currents with full-scale deflections between about 5 and 100 mA. For larger direct currents, up to about 20 A, the moving coil is shunted by a low resistance (Fig. 6.28(b)). Larger values of alternating currents can be determined by using a current transformer with the ammeter. The instrument can be converted to a voltmeter by the addition of a series resistor and can be used for both direct and alternating voltages up to about 2 kHz, giving r.m.s. values regardless of waveform. The instrument has a non-linear scale, is more expensive than a moving coil instrument and has a higher power consumption. One of the main uses of the instrument is as a wattmeter (see Chapter 10). Stray magnetic fields can affect the operation of the instrument and therefore the coil system is enclosed in a magnetic shield.

**Problems**

1  In repairing a moving coil meter the springs are replaced by ones which require twice the torque at any deflection. How will this affect the calibration of the meter?
2  Calculate the values of the shunts and multipliers required to convert a moving coil meter with a full-scale deflection of 1.0 mA and resistance of 50 Ω into:
   (a)  A milliammeter with a full-scale deflection of 200 mA.
   (b)  An ammeter with a full-scale deflection of 5 A.
   (c)  A voltmeter with a full-scale deflection of 1.0 V,
   (d)  A voltmeter with a full-scale deflection of 20 V.
3  Design a universal (Ayrton) shunt to provide an ammeter with direct current ranges of 100 mA, 1.0 A and 10 A. The moving coil meter to be used has a resistance of 50 Ω and gives a full-scale deflection with 1.0 mA.
4  A meter has a resistance of 1 kΩ and gives a full-scale deflection with 50 μA. What are the resistances required for a universal (Ayrton) shunt to provide current ranges of 10 mA, 100 mA and 1 A?
5  Design a circuit that can be used to provide a multi-range voltmeter with direct voltage ranges of 1.0, 10 and 100 V. The moving coil meter to be used has a resistance of 50 Ω and gives a full-scale deflection with 1.0 mA.
6  A moving coil voltmeter with a movement plus multiplier resistance of 10 kΩ gives a full-scale reading of 1 V. What is its sensitivity?
7  A moving coil voltmeter has a sensitivity of 20 000 Ω/V and is

used on its 50 V scale for the measurement of a voltage in a circuit having a Thévenin equivalent resistance of 30 kΩ. What will be the percentage loading error produced?

8   A moving coil voltmeter uses a 50 μA meter movement and is to be used on its 100 V scale to measure the voltage across a 2 MΩ resistor in a circuit which has a total resistance of 4 MΩ. What will be the percentage error in the reading as a consequence of loading?

9   A series type ohmmeter uses a moving coil meter with a full-scale current of 1.0 mA and resistance 50 Ω with a battery of e.m.f. 2.5 V. What will be the values of the resistance used to give full-scale deflection, i.e., the zero resistance reading, and the meter shunt required if the mid-scale resistance reading is to be 500 Ω?

10  Explain how the readings of average-responding, peak-responding and r.m.s.-responding electronic voltmeters will differ for a sinusoidal signal input.

11  A moving iron meter gives a full-scale deflection with 10 A. What will be the current for the quarter-scale and the mid-scale readings?

12  Compare the performance characteristics of moving coil, moving iron and thermocouple meters.

13  Explain why a.c. meters can be affected by a waveform error.

14  Suggest, giving reasons, meters which can be used for the measurement of (a) high frequency currents, (b) alternating current of about 1 A in a situation where high accuracy is not required but cheapness is essential, (c) a d.c. current to an accuracy of about ±1%, (d) an alternating current of about 1 mA at 50 Hz with an accuracy of about ±5%.

# 7 Digital meters

## Introduction

Digital meters give readings in the form of digits, i.e., a number, rather than in terms of the position of some pointer along a scale. The advantages of this are a reduction in the human reading error, faster reading of the value, and no parallax error. Because most of the quantities to be measured are analogue in nature, they need to be converted to a digital signal before they can be displayed by the instrument. Thus an analogue-to-digital converter (ADC) is usually an integral part of a digital instrument. There is thus in this chapter a consideration of the basic forms of analogue-to-digital converters in relation to digital voltmeters, the prime focus of the chapter. Digital counters are discussed in Chapter 12.

## The digital voltmeter

The digital voltmeter can be considered to be basically just an analogue-to-digital converter connected to a counter and a display unit (Fig. 7.1). The voltage to be measured, an analogue quantity, is sampled at some instant of time and converted by the ADC to a digital signal, i.e., a series of pulses with the number of the pulses being related to the size of the analogue voltage. These pulses are counted by a counter (see Chapter 12) and displayed as a series of digits.

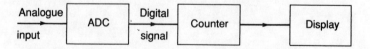

**Fig. 7.1**  Digital voltmeter principle

## Analogue-to-digital conversion

Analogue-to-digital converters (ADCs) convert an analogue signal with a size/amplitude which can vary continuously into a digit form which has a discrete number of levels. It is like taking a continuously varying height and converting it into a position up a ladder or staircase, the height then becoming so many rungs or steps. An analogue height of, say, 5 metres might be represented by a rung digit of 24. Each rung digit represents a movement of 1 rung. Most digital data formats used with electronic instrumentation are based on signal levels that are restricted to two states, i.e., *binary* values represented by the symbols 0 and 1. The binary digits of 0 and 1 are referred to as *bits*, a group of bits being referred to as a *word*. Thus, for example, a word might be 0101, such a word containing 4 bits. The position of the bits in a word has the significance that the least significant bit (LSB) is on the right end of the word and the most significant bit (MSB) on the left end. The denary value of the bits in a word is

$$2^{N-1} \ldots 2^4 \, 2^3 \, 2^2 \, 2^1 \, 2^0$$

MSB                  LSB

Thus 0101 is the denary number obtained by adding together $2^1$ and $2^3$, i.e., 10. The number of levels into which an analogue signal can be subdivided and so specified is determined by the number of bits in the word. Thus if we have $N$ bits there are $2^N$ levels. Thus a 4 bit word has $2^4$ levels, i.e., 16 levels.

Figure 7.2 illustrates such a conversion from an analogue signal 0 to 1.5 V into a 4-bit word. With no input all the bits in the word are 0. When the input voltage rises by 0.1 V then the first bit in the word becomes 1. When the input rises to 0.2 V then the first bit changes to 0 and the second bit becomes 1. Each rise in input of 0.1 V results in a bit being added to the word. The basic rules for adding binary numbers are:

$$0 + 0 = 0$$

$$0 + 1 = 1$$

$$1 + 1 = 10$$

With the 4-bit conversion the smallest change in input that will produce a change in binary output is 0.1 V. This is what is termed as the *resolution* of the converter. A change in input of less than 0.1 V produces no change in digital output. For a given range of input signal the bigger the word length of the converter the better its resolution. A 4-bit word length means the signal range is broken down into 16 levels, a 6-bit word into 64 levels and an 8-bit word into 256 levels. Analogue-to-digital converters typically have word lengths of 8, 10, 12, 14 or 16 bits.

The term *conversion time* is used to specify the time it takes a converter to generate a complete digital word when supplied with the analogue input. Typically, conversion times are of the order of microseconds. With the example given in Fig. 7.2 an analogue voltage of 1.0 V gives a digital output of 1010. There

| Input in V | Word | | | | Signal |
|---|---|---|---|---|---|
| 0.0 | 0 | 0 | 0 | 0 | |
| 0.1 | 0 | 0 | 0 | 1 | |
| 0.2 | 0 | 0 | 1 | 0 | |
| 0.3 | 0 | 0 | 1 | 1 | |
| 0.4 | 0 | 1 | 0 | 0 | |
| 0.5 | 0 | 1 | 0 | 1 | |
| 0.6 | 0 | 1 | 1 | 0 | |
| 0.7 | 0 | 1 | 1 | 1 | |
| 0.8 | 1 | 0 | 0 | 0 | |
| 0.9 | 1 | 0 | 0 | 1 | |
| 1.0 | 1 | 0 | 1 | 0 | |
| 1.1 | 1 | 0 | 1 | 1 | |
| 1.2 | 1 | 1 | 0 | 0 | |
| 1.3 | 1 | 1 | 0 | 1 | |
| 1.4 | 1 | 1 | 1 | 0 | |
| 1.5 | 1 | 1 | 1 | 1 | |

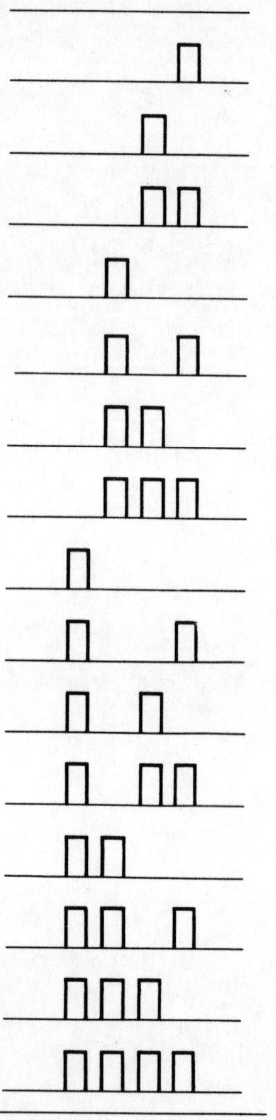

**Fig. 7.2** Analogue-to-digital conversion

is then no change in output until the analogue voltage has risen to 1.1 V when the output becomes 1011. Thus an output of 1010 can only mean that the voltage is between 1.0 and 1.1 V. The term *quantization error* is used for this uncertainty in the conversion, being the error due to the conversion of 1 bit.

### Example 1

What is the resolution of an analogue-to-digital converter having a word length of 10 bits if the analogue signal input range is 10 V?

*Answer*

The number of levels with a 10-bit word is $2^{10} = 1024$. Thus the resolution is $10/1024 = 9.8\,\text{mV}$.

**Analogue-to-digital converters**

There are a number of forms of analogue-to-digital converter, the most common being successive approximations, flash, ramp, dual ramp and voltage-to-frequency. The successive approximations, flash and ramp forms are examples of what can be termed sampling ADCs in that they provide digital values equivalent to the voltage at the instant of time at which the signal is sampled. The dual ramp and voltage-to-frequency forms are examples of integrating ADCs in that the average value of the voltage is indicated over a fixed measurement time. Integrating forms take longer to carry out a measurement but have better noise rejection.

With the *successive approximations* form a sample of the analogue input voltage is taken and then compared with a voltage which is increased in increments until its total value reaches the input voltage (Fig. 7.3). This incrementally increasing voltage is produced by a clock emitting a regular sequence of pulses which are counted and converted into an analogue signal by a digital-to-analogue converter. The resulting analogue signal is compared with the input voltage and when it rises to the level above it, the pulses from the clock are stopped from reaching the counter and thus the counter reading is the digital equivalent of the analogue input voltage. Typically, an 8-bit meter will have a conversion time of about $10\,\mu\text{s}$; thus sampling times are typically of the order of 1000 times per second or more. The successive approximations form of digital voltmeter is one of the faster responding voltmeters. The method outlined above is rather like weighing an object on a laboratory balance by building up to the balance weight by putting in the balance pans an increasing number of the smallest size weights. A faster weighing method is to start with a large weight, check for balance and if too large, to try half the weight. This is continued with the weights being halved each time until one is found which is too light. That weight is then retained and then the process of adding half size weights continued until the smallest weight is reached. A similar procedure can be used with the ADC. Thus for fast responses, instead of comparing the input voltage with steadily mounting voltage increments and building up to the required voltage, the comparison is made first with the analogue of the most significant bit, then the second most significant bit, in much the same way as the faster weighing method. The sequence might thus be: try 1000; if too large, try 0100. If too small, try 0110. If too small, try 0111. If too large,

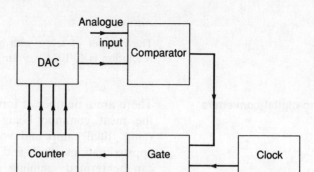

**Fig. 7.3**  Successive approximations ADC

then the result is 0110. Because each of the bits in the word is tried in sequence it takes only $N$ steps for a $N$-bit word. If the clock has a frequency of $f$, the smallest time in which a change can be made to a word is $N\tau$ where $\tau$ is the time between successive pulses, i.e., $1/f$. The conversion time is thus $N/f$.

The term *flash converter* is used for a very fast form of ADC where simultaneous comparisons are made between the analogue signal and reference signals. For an $N$-bit conversion $2^{N-1}$ comparators are used, one for each digit. Such an ADC can have a conversion time of the order of 10 ns.

The *ramp* form of ADC (Fig. 7.4) gives the simplest and cheapest form of digital voltmeter. The input analogue voltage is applied to a comparator and the time taken for a ramp voltage to climb from 0 V to the analogue voltage measured. This time is obtained in digital form by counting the number of pulses produced by a clock during the time the gate is open, the gate being opened when the ramp starts and closed when the ramp and analogue voltages are equal. Because of non-linearities in the shape of the ramp waveform and its lack of noise rejection, accuracy is typically limited to about ±0.05%. Sample rates can be up to about 1000 times per second.

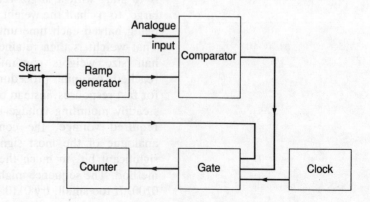

**Fig. 7.4**  Ramp ADC

The *dual ramp* form (Fig. 7.5) has the input analogue voltage integrated over a time $t_1$ equal to one cycle of the line frequency. For a constant analogue voltage input, a steadily increasing voltage is produced across the capacitor of the integrator. At the end of the time a switch operates and disconnects the analogue voltage from the integrator, leaving the capacitor with a charge. Since $i = dq/dt = q/t_1$ for a constant rate of charging, the charge $Q_{in}$ on the capacitor is $it_1 = v_{in}t_1/R$. Then a constant negative analogue reference voltage $v_{ref}$ is switched to the integrator input. The time $t_2$ is then measured for the integrator output to reach zero, i.e., for $v_{ref}t_2/R$ to equal $v_{in}t_1/R$ (Fig. 7.6). Thus the time $t_2$ is a measure of $v_{in}$. This form of ADC has the advantage of noise and line-frequency signal rejection but since it integrates the signal over one cycle of the mains frequency it has a conversion time of only the reciprocal of the mains frequency, i.e., 1/50 or 1/60s. Accuracy is about $\pm 0.005\%$.

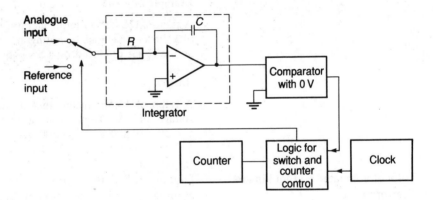

**Fig. 7.5**  Dual ramp ADC

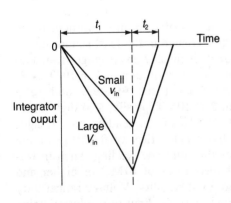

**Fig. 7.6**  The dual ramp

With the *voltage-to-frequency* form of ADC the input analogue voltage is converted into a set of pulses whose frequency is proportional to the size of the input voltage. The frequency is then determined by counting the number of pulses occurring in some fixed time interval. Figure 7.7(*a*) shows a block diagram of this form of ADC and Fig. 7.7(*b*) the form that can be taken by the voltage-to-frequency element. An integrator is used to integrate the analogue input voltage $v_{in}$ over the time it takes for the integrator output to go from 0 to $-v_{ref}$. This time is $RCv_{ref}/v_{in}$. At the end of this time, a pulse generator emits a single pulse. Since the input to the integrator is still present, the process repeats itself and the result is a sequence of pulses with a frequency determined by $v_{in}$.

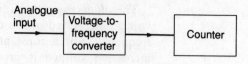

(a)

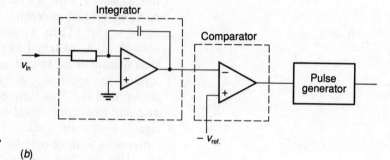

(b)

**Fig. 7.7**(a) Voltage-to-frequency ADC,
(b) voltage-to-frequency converter

### Example 2

A ramp ADC has a ramp generator which gives a ramp with a slope of 1 V/ms and a clock operating at a frequency of 100 kHz. What will be the count produced by a 5 V input signal?

*Answer*

A ramp of 1 V/ms means that it will take 5 ms to reach the input voltage of 5 V. Hence during that time the clock will produce $5 \times 10^{-3} \times 100 \times 10^3 = 500$ pulses.

## Digital voltmeter and multimeter performance

Digital voltmeters provide a numerical readout which eliminates the interpolation and parallax errors associated with the reading of analogue displays. Displays are in denary form and are generally between $3\frac{1}{2}$ and $8\frac{1}{2}$ digits, the half being used in the specification because the most significant digit can only take the value 0 or 1, all other digits being able to take values between 0 and 9. The resolution of such an instrument is the voltage change which gives a change in the least significant bit of the meter display. A $3\frac{1}{2}$ digit display has a range from 1 to 1999 in its display and thus has a resolution of 1 in 1999. A $8\frac{1}{2}$ digit display has a range from 1 to 19 999 999 and so a resolution of virtually 1 in $2 \times 10^8$. Typically, a $3\frac{1}{2}$ digit meter will have an accuracy of ±0.1% of the reading plus 1 digit, while a $8\frac{1}{2}$ digit display has an accuracy of 0.0001% of the reading plus 0.000 03% of the full-scale reading. Such instruments usually have input resistances of 10 MΩ or higher and capacitances of 40 pF, and good stability. Voltage ranges vary from about 100 mV to 1000 V with the limit of resolution being about 1 μV. Depending on the form of the ADC so the meter will sample the analogue voltage or integrate it over a fixed time. As an example of sampling voltmeter, a successive approximations form will typically have a conversion time of about 10 μs, with a flash conversion form being about 10 ns. The smaller the conversion time, the more an instrument can respond to a changing input or sudden peaks. Where such situations are anticipated, an analogue meter may be preferred. An example of an integrating voltmeter is the dual ramp type, such a form integrating over the line frequency and so giving 25 or 30 conversions per second.

The basic digital voltmeter is a d.c. meter. It can be used for the measurement of alternating voltage by using a rectification circuit, in a similar way to permanent magnet moving coil instruments (see Chapter 6). Such rectification methods give average values and since the instruments are generally scaled to read r.m.s. values corrections have to be made for non-sinusoidal waveforms. Accuracy typically varies from about $\pm1\%$ of the reading plus three digits with a $3\frac{1}{2}$ digit display to $\pm0.05\%$ of the reading plus 0.03% of the full-scale reading for a $8\frac{1}{2}$ digit display. The frequency range varies from about 45 Hz to 10 kHz for a $3\frac{1}{2}$ digit display to 10 Hz to 100 kHz for a $8\frac{1}{2}$ digit display. Voltage ranges vary from full scale readings of about 100 mV to 1000 V r.m.s. with an input impedance of about 10 MΩ with 100 pF.

Current, d.c. or a.c., can be measured by the digital voltmeter being used to measure the potential difference across a standard resistor. Accuracy is typically about $\pm0.2\%$ of the reading plus two digits for d.c. and $\pm1\%$ of the reading plus two or more digits for a.c. For both d.c. and a.c. the ranges are from about 200 μA to 2 A and the voltage drop across the instrument less than 0.3 V. The frequency range is about 45 Hz to 1 kHz.

Resistances can be measured using a digital voltmeter by passing a known current through the unknown resistance and then using the voltmeter to measure the resulting potential difference across it. High precision instruments, however, use a different method. A current is passed through a standard resistor and the unknown resistor and the potential differences across the two compared. Because the current is the same through both, the potential difference ratio is the ratio of the resistances. Accuracy varies from about $\pm0.1\%$ of the reading plus one digit for a $3\frac{1}{2}$ digit meter to $\pm0.0002\%$ of the reading plus $\pm0.0004\%$ of the full-scale reading for a $8\frac{1}{2}$ digit display. The resistances ranges are from about 200 Ω to 1000 MΩ.

Figure 7.8 shows a block diagram of a digital multimeter which can be used for the measurement of d.c. and a.c. voltage and current, and resistance. The methods used for the various measurements are as outlined above. Some digital

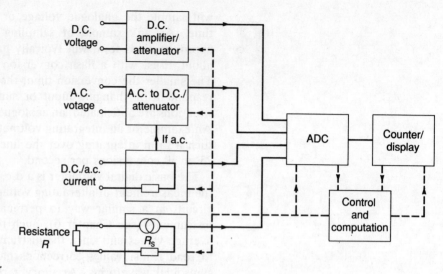

**Fig. 7.8** Digital multimeter

meters can give a digital output signal which can be fed to a data logger or computer.

**Problems**

1　What are the main advantages digital instruments have over analogue instruments?

2　What is the resolution of an analogue-to-digital converter having a word length of 12 bits if the analogue signal input range is 100 V?

3　What is the percentage resolution of a 10-bit ADC?

4　Compare the performances of the following forms of analogue-to-digital converters: successive approximations, flash, ramp, dual ramp, voltage-to-frequency.

5　A ramp ADC has a ramp generator which gives a ramp with a slope of 1 V/ms and a clock operating at a frequency of 50 kHz. What will be the count produced by a 10 V input signal?

6　What is the conversion time of an 8-bit successive approximations ADC if it has a clock operating at 1 MHz?

7　For the following situations suggest the form of ADC likely to be required for the digital voltmeters:

(*a*)　A cheap instrument of general use where high accuracy and fast conversion times are not required.

(*b*)　A voltmeter for use where signals are changing very rapidly.

(*c*)　A voltmeter which can be used with minimum error in a situation where there is likely to be a lot of line frequency interference.

8　A specification for a digital voltmeter states that it is a $3\frac{1}{2}$ digit meter. What does this mean?

9　What is the smallest voltage change that can be determined by a $3\frac{1}{2}$ digit voltmeter with a range of 0 to 100 V?

# 8 Recorders

**Introduction**

The wide range of elements used for the presentation of data can be broadly classified into two groups: indicators and recorders. *Indicators* give an instant visual indication of the process variable, while *recorders* record the output signal over a period of time and automatically give a permanent record. Both indicators and recorders can be subdivided into two groups of devices, *analogue* and *digital*. With analogue chart recorders the input signal is translated into some position on a paper chart and a marking mechanism marks the point. By moving either the marking mechanism or the chart in a controlled way with time, an analogue record can be obtained of how the input signal varied with time. The two main forms of analogue chart recorder are the galvanometric type and the potentiometric type. With digital recording involving paper charts, the output is in the form of a print-out with digits. Another important method of recording, for both analogue and digital inputs, involves the use of magnetic tape or disks.

**Direct reading recorders**

The *direct reading type* of chart recorder has a pen which is moved over a circular chart by the direct action of the quantity being measured. For example, with a pressure recorder the displacement of the end of a Bourdon tube or bellows as a result of a change in pressure may be used to move the pointer across the chart (Fig. 8.1). With a temperature recorder the displacement of the end of a bimetallic strip may be used. Such recorders have circular charts which rotate at a constant rate, usually one revolution in 12 hours, 24 hours or 7 days. The chart rotates at a constant rate and thus equal angles are covered in equal intervals of time. This means that the distance moved by the pen in equal intervals of time depends on its distance out from the chart centre. Another complicating factor is that the tip of the pen moves outwards from the chart centre in curved radial lines because the movement of the pen

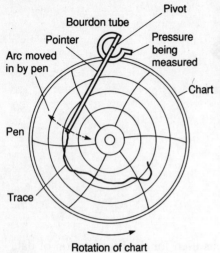

**Fig. 8.1** Direct-reading type of chart recorder

is in the arc or a circle. The resulting chart is thus difficult to read and interpolation is difficult. There are particular problems in determining values for traces close to the centre of the chart where the radial lines are very close together. Simultaneous recording of up to four separate variables is possible, the accuracy being generally about ±0.5% of the full-scale deflection of the signal.

### Galvanometric recorders

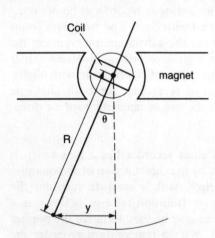

**Fig. 8.2** Galvanometric type of chart recorder

The *galvanometric type* of chart recorder (Fig. 8.2) works on the same principle as the moving coil meter movement described in Chapter 6. It consists of a coil suspended in the magnetic field of a permanent magnet. When a current passes through the coil, the coil experiences a torque. The main difference is that usually the torque opposing the rotation of the coil is produced by the twisting of the suspension of the coil rather than by springs. In its simplest form, a pointer with a pen at its end is attached to the coil and rotates with it. Movement of the pointer as a result of a current through the coil results in an ink trace being produced on a chart.

If $R$ is the length of the pointer and $\theta$ the angular deflection of the coil, the displacement $y$ of the pen is

$$y = R \sin \theta \qquad [1]$$

Since the angle $\theta$ rotated is proportional to the current through the coil (see Chapter 6), the current is proportional to $\sin^{-1} y/R$. This is a non-linear relationship. If the angular deflections are restricted to less than ±10° then $\sin \theta$ is approximately $\theta$ and so the current is reasonably proportional to $y/R$. For such angles the error due to non-linearity is less than 0.5%. A greater problem is, however, the fact that the pen moves in an arc rather than a straight line. Thus curvilinear paper (Fig. 8.3) is used for the plotting. However, there are difficulties in interpolation for points between the curved lines.

An alternative form of 'pen' chart recorder which leads to rectilinear charts rather than curvilinear charts is the *knife*

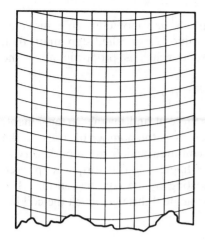

**Fig. 8.3** Curvilinear chart paper

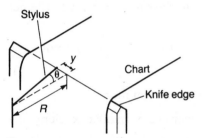

**Fig. 8.4** Knife edge recorder

*edge recorder*. One version of this uses a heated stylus, instead of an ink pen, and heat-sensitive paper which moves over a knife edge (Fig. 8.4). The paper may be impregnated with a chemical that shows a marked colour change when heated by contact with the stylus or the stylus burns away temperature-sensitive outer layers which coat the paper. The use of the knife edge avoids the curved trace but the non-linear relationship between $\theta$ and displacement $y$ still exists. The length of trace $y$ produced on the paper by a deflection $\theta$ is

$$y = R\tan\theta \qquad [2]$$

The non-linearity error is slightly greater than for the pen form of recorder. If deflections are restricted to less than $\pm 10°$ the error is less than 1%.

Usually with the pen or knife edge form of galvanometric recorder there is some electronic amplification of the input signal. This leads to sensitivities which are generally of the order of 1 cm of pen displacement per mV, input resistances of about $10\,\mathrm{k}\Omega$, a bandwidth from d.c. to about 50 Hz and accuracies of the order of $\pm 2\%$ of the full-scale deflection.

## Ultraviolet galvanometer recorder

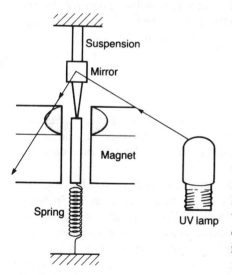

**Fig. 8.5** Ultraviolet galvanometric recorder

There are a number of ways by which the movement of the coil in a galvanometric recorder may be transformed into a trace on a chart. An alternative to the pen form is the *ultraviolet recorder* which involves a small mirror attached to the suspension (Fig. 8.5). A narrow beam of ultraviolet light is directed at the mirror and when the coil rotates the reflected beam is swept across the chart. The chart uses photosensitive paper and so a trace is produced when it is developed.

The use of an 'optical pointer' rather than a material pointer enables longer 'pointer' lengths to be used and so greater sensitivity. With a bandwidth of d.c. to about 50 Hz the sensitivity is typically about 5 cm/mV, the coil having a resistance of about $80\,\Omega$. Higher bandwidth instruments have lower sensitivities, e.g., a bandwidth extending up to 5 kHz with a sensitivity of about 0.0015 cm/mV and a coil resistance of about $40\,\Omega$. The limiting frequency for this type of instrument is about 13 kHz. Typically the accuracy is about $\pm 2\%$ of the full-scale deflection. Since optical pointers can cross each other without interference, it is quite common to have 6, 12 or 25 galvanometer mountings side by side in the same magnet block and so enable simultaneous recordings of many variables to be made.

**Dynamic behaviour of galvanometric recorders**

When the current through a galvanometer coil suddenly changes from zero to some value $I$, a so-called step input, then the coil experiences a torque which is proportional to the current:

torque due to current $I = K_c I$

where $K_c$ is a constant for the coil and magnet arrangement. $K_c$ has the value $NAB$, where $N$ is the number of turns on the coil, $A$ the cross-sectional coil area and $B$ the flux density at right angles to the coil (see Chapter 6). This torque is opposed by a torque due to the twisting of the coil suspension. The torque is proportional to the angle $\theta$ through which the coil has rotated:

opposing torque $= K_s \theta$

where $K_s$ is a constant related to the suspension used. The net torque acting on the coil is thus

net torque = torque due to current − opposing torque

$$= K_c I - K_s \theta \qquad [3]$$

This net torque gives an angular acceleration:

net torque = moment of inertia $J$ × angular acceleration

For angular motion the moment of inertia is the equivalent of mass in linear motion. Hence

$$J \times \text{angular acceleration} = K_c I - K_s \theta \qquad [4]$$

The circuit in which the current is being measured by the recorder can be considered in terms of the Thévenin equivalent circuit (Fig. 8.6). The current $I$ is thus

$$I = \frac{V_{\text{Th}}}{R_{\text{Th}} + R_r} \qquad [5]$$

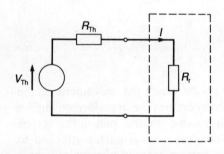

**Fig. 8.6** Thévenin equivalent circuit

where $V_{\text{Th}}$ is the Thévenin equivalent voltage, $R_{\text{Th}}$ its effective resistance and $R_r$ the effective resistance of the recorder circuit in which the galvanometer coil is. This equation, however, needs to be modified during the motion of the galvanometer coil.

The current through the coil causes it to rotate. But the rotation of a coil in a magnetic field results in an induced e.m.f. being produced in the coil. For a length $L$ of conductor (Fig. 8.7) moving with a linear velocity $v$ at right angles to a magnetic field of flux density $B$ the induced e.m.f. $E$ is

$$E = BLv$$

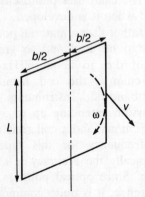

**Fig. 8.7** Moving coil

Such a conductor is the vertical side of the galvanometer coil. The vertical side is moving in a circular path of radius $\frac{1}{2}b$, where $b$ is the breadth of the coil. The angular velocity $\omega$ of such a rotation is related to the linear velocity $v$ by $v = r\omega$ where $r$ is the radius of the path. Hence $v = \frac{1}{2}b\omega$. Thus $E = \frac{1}{2}BLb\omega$. Since there are two sides and $N$ turns $E = NBLb\omega$. Since the coil area $A = Lb$ then

$$E = NBA\omega = K_c\omega$$

Thus the voltage in the circuit is $(V_{Th} - K_c\omega)$. Thus equation [5] becomes modified during the coil motion to become

$$I = \frac{V_{Th} - K_c\omega}{R_{Th} + R_r}$$

Hence substituting for $I$ in equation [4]

$$J \times \text{angular acceleration} = \frac{K_c(V_{Th} - K_c\omega)}{R_{Th} + R_r} - K_s\theta \qquad [6]$$

The angular velocity is the rate at which the angle changes with time and can be written as $d\theta/dt$. The angular acceleration is the rate at which the angular velocity changes with time and can be written as $d^2\theta/dt^2$. Hence

$$\frac{d^2\theta}{dt^2} + \frac{K_c^2}{J(R_{Th} + R_r)}\frac{d\theta}{dt} + \frac{K_s\theta}{J} = \frac{K_cV_{Th}}{J(R_{Th} + R_r)} \qquad [7]$$

This is a second-order differential equation and describes the motion of the galvanometer coil.

When the recorder galvanometer stops moving and reaches the steady reading there is no angular acceleration or angular velocity and thus both $d^2\theta/dt^2$ and $d\theta/dt$ are zero. Hence

$$\frac{K_s\theta}{J} = \frac{K_cV_{Th}}{J(R_{Th} + R_r)}$$

The steady-state voltage sensitivity is thus

$$\text{voltage sensitivity} = \frac{\theta}{V_{Th}} = \frac{K_c}{K_s(R_{Th} + R_r)} \qquad [8]$$

The differential equation describes the behaviour of a system which has a natural frequency of oscillation $\omega_n$, where

$$\omega_n = 2\pi f_n = \sqrt{(K_s/J)} \qquad [9]$$

and is damped with a damping factor of

$$\text{damping factor} = \frac{K_c^2}{2(K_sJ)^{\frac{1}{2}}(R_{Th} + R_r)} \qquad [10]$$

The damping factor involves the effective resistance of the circuit connected to the recorder. Thus the damping can be altered by adding resistors in series or parallel with the recorder input.

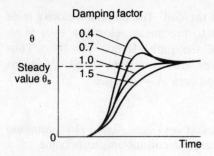

**Fig. 8.8** Response of the galvanometer to a step current input

**Table 8.1** Overshoot and damping factor

| Damping factor | % overshoot |
| --- | --- |
| 1.00 | 0.0 |
| 0.91 | 0.1 |
| 0.82 | 1.2 |
| 0.72 | 4.0 |
| 0.62 | 8.4 |
| 0.50 | 16.5 |
| 0.40 | 25.0 |

Figure 8.8 shows how the response of the galvanometer depends on the damping factor when there is a step current input. The *damping factor* expresses the damping as a fraction of that which gives critical damping, i.e., critical damping has a damping factor of 1.0. Critical damping occurs when the galvanometer coil responds to the step input by changing to the steady deflection without any overshoot in the minimum of time. In Fig. 8.8 it is the deflection indicated for the damping factor of 1.0. At damping factors of less than 1.0 the deflection overshoots the steady value before settling back to it; the coil is then said to be under-damped. With a damping factor of 0.4 the percentage overshoot is about 25%, at 0.7 it is about 5%. Table 8.1 shows the overshoot for other values of damping. With damping factors greater than 1.0 the coil just takes a long time to attain the steady value, the coil then being said to be over-damped.

The response of the galvanometer to different frequency inputs depends on the value of the natural frequency and the damping. Figure 8.9 shows the frequency response. The frequency band for which the steady deflection value $\theta_s$ is attained depends on the damping. The maximum frequency for which this occurs is when the damping factor is of the order of 0.7 and is almost the natural frequency.

Galvanometers for use at high frequencies thus require a damping factor of about 0.7 and a high natural frequency. One way by which the natural frequency can be increased is by reducing the moment of inertia $J$ of the coil. This can be achieved by using a slim coil, i.e., one with a small breadth. However, reducing the moment of inertia also affects the damping factor, a decrease in moment of inertia increasing the damping factor. Another way by which the natural frequency

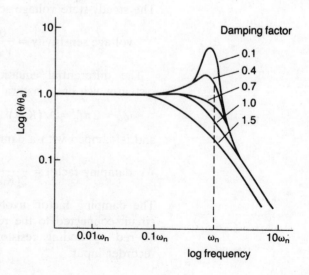

**Fig. 8.9** Frequency response of a recorder galvanometer

can be increased is by increasing $K_s$. This, however, also affects the sensitivity of the galvanometer. Thus galvanometers designed for use at high frequencies tend to have low sensitivities.

**Example 1**

How is the damping factor and steady state voltage sensitivity of a recorder galvanometer affected by a resistance of $100\,\Omega$ being included in series with the input circuit, if the damping factor is 0.6 without it? The recorder has a resistance of $60\,\Omega$ and the circuit connected to it a Thévenin resistance of $160\,\Omega$.

*Answer*

Equation [10] gives

$$\text{damping factor} = \frac{K_c^2}{2(K_sJ)^{\frac{1}{2}}(R_{\text{Th}} + R_{\text{r}})}$$

Hence, without the extra $100\,\Omega$,

$$0.6 = \frac{K_c^2}{2(K_sJ)^{\frac{1}{2}}(160 + 60)}$$

With the extra $100\,\Omega$,

$$\text{damping factor} = \frac{K_c^2}{2(K_sJ)^{\frac{1}{2}}(260 + 60)}$$

Hence, dividing these two equations,

$$\frac{\text{damping factor}}{0.6} = \frac{160 + 60}{260 + 60}$$

The new damping factor is thus 0.4. The result of this is that the galvanometer will give considerably more overshoot.

The steady-state voltage sensitivity is given by equation [8] as

$$\text{sensitivity} = \frac{\theta}{V_{\text{Th}}} = \frac{K_c}{K_s(R_{\text{Th}} + R_{\text{r}})}$$

Without the extra resistance,

$$\text{sensitivity without } R = \frac{K_c}{K_s(160 + 60)}$$

With the extra resistance

$$\text{sensitivity with } R = \frac{K_c}{K_s(260 + 60)}$$

Hence, dividing these two equations,

$$\frac{\text{sensitivity with } R}{\text{sensitivity without } R} = \frac{160 + 60}{260 + 60}$$

The voltage sensitivity is decreased by a factor of 0.7.

**Example 2**

How are the damping factor and the steady-state voltage sensitivity of a recorder galvanometer affected by a $100\,\Omega$ resistance being connected (a) in series with the input circuit and (b) in parallel with it? The galvanometer has a resistance of $80\,\Omega$ and the circuit a Thévenin resistance of $100\,\Omega$.

*Answer*

(a) The damping factor is given by equation [10] as

$$\text{damping factor} = \frac{K_c^2}{2(K_sJ)^{\frac{1}{2}}(R_{Th} + R_r)}$$

Hence for the resistance in series

$$\frac{\text{new damping factor}}{\text{initial damping factor}} = \frac{R_{Th} + R_r}{R + R_{Th} + R_r} = \frac{180}{280}$$

The new damping factor is 0.64 times the initial damping factor.
  The steady-state voltage sensitivity is given by equation [8] as

$$\text{sensitivity} = \frac{\theta}{V_{Th}} = \frac{K_c}{K_s(R_{Th} + R_r)}$$

Hence for the resistance in series

$$\frac{\text{new sensitivity}}{\text{initial sensitivity}} = \frac{R_{Th} + R_r}{R + R_{Th} + R_r} = \frac{180}{280}$$

The new sensitivity is 0.64 times the initial sensitivity.

(b) For the resistance in parallel the combined resistance of the meter and the Thévenin resistance of the input circuit becomes

$$\frac{1}{R} = \frac{1}{100} + \frac{1}{100}$$

and so is $50\,\Omega$. Hence

$$\frac{\text{new damping factor}}{\text{initial damping factor}} = \frac{280}{130}$$

The new damping factor is 2.2 times the initial damping factor.
  In considering the voltage sensitivity account has to be taken of not only the change in the effective resistance but also the voltage applied to the recorder. The Thévenin equivalent voltage for the input circuit is reduced by a factor of $R_p/(R_p + R_{Th})$, where $R_p$ is the resistance inserted in parallel. Thus in this case the Thévenin equivalent voltage is reduced by a factor of $100/(100 + 100)$, i.e., by a half. Hence, since

$$\text{voltage sensitivity} = \frac{\theta}{V_{Th}} = \frac{K_c}{K_s(R_{Th} + R_r)}$$

with the parallel resistor,

$$\frac{\theta}{0.5V_{Th}} = \frac{K_c}{K_s(50 + 80)}$$

and so the new sensitivity is

$$\text{new sensitivity} = \frac{\theta}{V_{\text{Th}}} = \frac{0.5K_{\text{c}}}{K_{\text{s}} \times 130}$$

Since the initial sensitivity was

$$\text{initial sensitivity} = \frac{K_{\text{c}}}{K_{\text{s}}(100 + 80)}$$

then the new sensitivity is $0.5 \times 180/130 = 0.69$ times the initial sensitivity.

**Potentiometric recorders**

The *potentiometric recorder*, or, as it is sometimes referred to, the *closed-loop recorder* or *closed-loop servo recorder*, is a self-balancing potentiometer. The output from a potentiometer is automatically adjusted until it matches the input voltage to the recorder. This adjustment is by the movement of a sliding contact along the potentiometer track. Since the sliding contact is linked mechanically to a pen then the movement of the slider to achieve a matching of the voltages results in the pen moving to a position on a chart related to the size of the voltage. Figure 8.10 shows the basic arrangement. The position of the pen is monitored by means of a slider which moves along a linear potentiometer. The position of the slider determines the potential applied to an operational amplifier. The amplifier subtracts the measurement signal from the signal from the transducer. The output from the amplifier is thus a signal related to the difference between the pen and transducer signals. This signal is used to operate a servo motor which in turn controls the movement of the pen across the chart. The pen thus ends up moving to a position where the result is no difference between the pen and transducer signals.

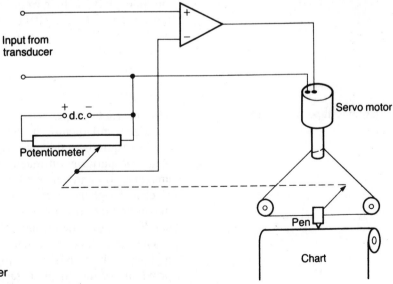

**Fig. 8.10** Potentiometric recorder

Potentiometric recorders typically have high input resistances and higher accuracies (about ±0.1% of full-scale reading) than galvanometric recorders, but considerably slower response times. Response times are typically of the order of 1–2 s and so the bandwidth is only d.c. to 1 or 2 Hz. They can thus only really be used for d.c. or slowly changing signals. Because of friction there is a minimum current required to get the motor operating. There is thus some error due to the recorder not responding to a small transducer signal. This error is known as the *dead-band*. Typically it is about ±0.3% of the range of the instrument. Thus for a range of 5 mV the dead-band error amounts to ±0.015 mV.

### X-Y recorder

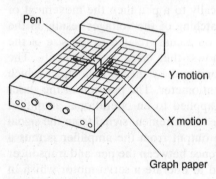

**Fig. 8.11** *X-Y recorder*

*X-Y* recorders are a form of potentiometric recorder with the pen driven by two motors, one motor controlling the movement in the *X*-direction and the other in the *Y*-direction (Fig. 8.11). This means that one variable can be applied to the *X* input and the other to the *Y* input and the recorder will plot a graph showing how one of the variables changes with respect to the other. Most *X-Y* recorders contain a timebase signal which can be internally applied to the *X* input and so enable a graph of the variation with time of the *Y* input to be plotted. Typically they have input ranges from between 0.25 mV/cm to 10 V/cm with an accuracy of about ±0.1% of the full-scale reading. The internal timebase is usually between about 0.25 and 50 s/cm with an accuracy of about ±1%. They have a dead-band of about 0.1% of the full-scale reading. Like the potentiometric recorder, because the pen is moved mechanically they are not able to track rapidly changing variables and so their frequency response is low, typically no more than about 5 or 10 Hz. The term *slewing rate* or speed is used to specify the maximum velocity with which the pen can be moved after its acceleration has stopped. This is typically about 10 cm/s, though higher speed recorders are possible.

### Magnetic tape recorders

The magnetic tape recorder can be used to record both analogue and digital signals. It consists of a recording head which responds to the input signal and produces corresponding magnetic patterns on magnetic tape, a replay head to do the converse job and convert the magnetic patterns on the tape to electrical signals, a tape transport system which moves the magnetic tape in a controlled way under the heads, and signal conditioning elements such as amplifiers and filters.

The recording head consists of a core of ferromagnetic material which has a non-magnetic gap (Fig. 8.12). The proximity of the magnetic tape to the non-magnetic gap means

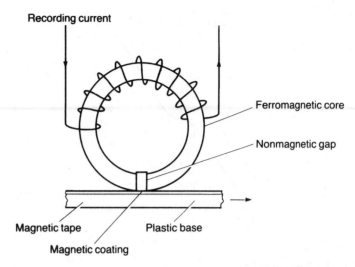

Recording current

Ferromagnetic core

Nonmagnetic gap

Magnetic tape

Plastic base

Magnetic coating

**Fig. 8.12** Basis of a magnetic recording head

that the magnetic circuit through which the magnetic flux 'flows' is the core and that part of the magnetic tape spanning the gap. Electrical signals are fed to a coil which is wound round the core and result in the production of magnetic flux in the magnetic circuit, and hence in that part of the magnetic tape spanning the gap. The magnetic tape is a flexible plastic base coated with a ferromagnetic powder. When there is magnetic flux passing through a region of the tape it becomes permanently magnetized, i.e., has a remanent flux density. Hence a magnetic record is produced of the electrical input signal.

The recording head and the replay heads have similar forms of construction. Thus when a piece of magnetized tape bridges the non-magnetic gap of a replay head, magnetic flux is induced in the core. Flux changes in the core induce e.m.f.s in the coil wound round the core. Thus the output from the coil is an electrical signal which is related to the magnetic record on the tape.

Figure 8.13(a) shows the relationship between the remanent flux density on the magnetic tape and the current in the coil wound round the recording or playback head. (Note: the horizontal axis is more often represented as $H$, the magnetizing force, with $H$ equal to $NI$ where $N$ is the number of turns and $I$ the current.) If the input signal has a current which fluctuates about the zero then the fluctuations in remanent magnetism produced on the tape are not directly proportional to the current. There is, in particular, distortion for very small currents. This can be reduced by adding a steady d.c. current to the signal, i.e., a d.c. bias. This shifts the signal to more linear parts of the graph. An alternative to this is to add a high frequency a.c. current to the signal and give an a.c. bias. The

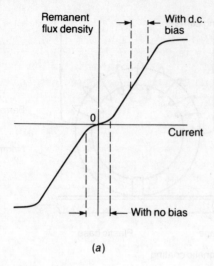

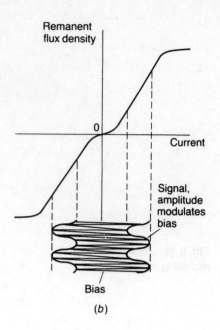

**Fig. 8.13** Effect of (a) d.c. bias, (b) a.c. bias

input signal is then used to modulate the amplitude of the a.c. signal and hence results in the signal being shifted to more linear parts of the graph (Fig. 8.13(b)).

If the input signal is sinusoidal with a frequency $f$ then a sinusoidal variation in magnetization is produced along the tape. The time interval for one cycle is $1/f$ and so if the tape is moving with a constant velocity $v$ then the distance along the tape taken by one cycle is $v/f$. This distance is called the *recorded wavelength*.

$$\text{recorded wavelength} = \frac{v}{f} \qquad [11]$$

The minimum size this recorded wavelength can have is the gap width. At this wavelength the average magnetic flux across the gap is the average of one cycle and this has a zero value. The size of the gap thus sets an upper limit to the frequency response of the recorder at a particular tape velocity. Typically tape velocities range between about 23 mm/s to 1500 mm/s with a gap width of 5 μm, hence upper frequency limit ranges from about 4.6 kHz up to 300 kHz.

The output from the replay head is proportional to the rate of change of flux $\Phi$ in the head core, i.e., $d\Phi/dt$. With a recording of a sinusoidal current input there is a sinusoidal variation of flux on the recording tape. Hence movement of the tape past the recording head will produce a sinusoidal variation of flux in the core. Thus

$$\Phi = \Phi_m \sin \omega t$$

$$\frac{d\Phi}{dt} = \omega\Phi_m \cos\omega t$$

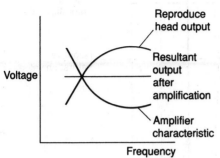

**Fig. 8.14**  Amplitude equalization

Thus the head output is proportional to $\omega\Phi_m \cos\omega t$, where $\omega = 2\pi f$, $f$ being the frequency, and $\Phi_m$ the maximum flux. Thus the head output depends not only on the flux recorded on the tape but also the frequency. To overcome this, amplitude equalization is used. An amplifier is used which has a transfer function which decreases as the frequency increases in such a way that feeding the replay head output through it results in a linear output (Fig. 8.14).

A consequence of the replay head output being proportional to the frequency is that for very low frequencies the output may be very small and so high equalization amplification is required. This will also amplify noise picked up by the replay head. A consequence of this is that there is a lower frequency limit for which the recorder can be used. This is generally about 100 Hz.

The above is a discussion of what is termed *direct recording*. With this form the input signal directly determines the magnetic flux recorded on the tape. An alternative to this is *frequency modulation*. With this the carrier frequency is varied in accordance with the fluctuations of the input signal. Thus the input signal is frequency modulated before entering the recording head. Because the carrier frequency is many kilohertz, the direct recording problem of dealing with low frequencies does not occur. Frequency modulation recording can thus be used down to 0 Hz, i.e., d.c. However, the upper frequency limit is less than with direct recording. Typically it is about a third of the carrier frequency, i.e., in the region 2–80 kHz. Frequency modulation tends to give a better signal-to-noise ratio than with direct recording. Particularly vital with frequency modulation is the need to keep the tape at a constant speed since fluctuations in this can lead to apparent frequency fluctuations.

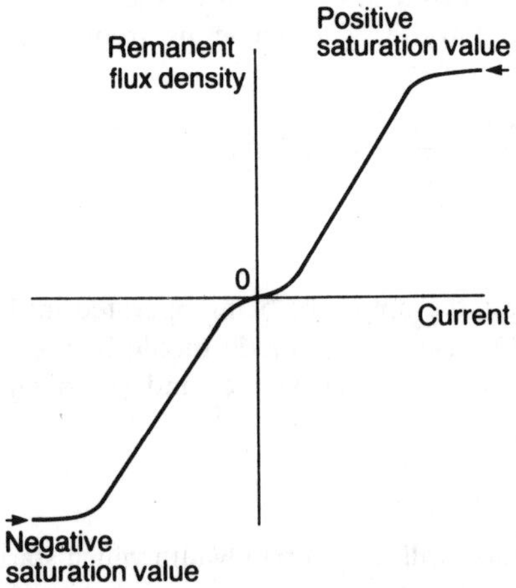

*Digital recording* involves the recording of signals as a coded combination of bits, i.e., a signal representing 0 or 1. A commonly used method is the *non-return to zero* (NRZ) method. With this system the flux recorded on the tape is either at the positive saturation value or the negative saturation value (Fig. 8.15). The method uses no change in flux to represent 0 and a change in flux 1. Figure 8.16 illustrates this for the number 0110101. Since the output from the replay head depends on the rate of change of flux on the tape, outputs only occur where the recorded tape has a change of flux. Thus the output is a pulse whenever a 1 was recorded. Digital recording has the advantages over analogue recording of higher accuracy and relative insensitivity to tape speed.

Recorders generally have more than one recording head.

**Fig. 8.15**  Saturation flux values

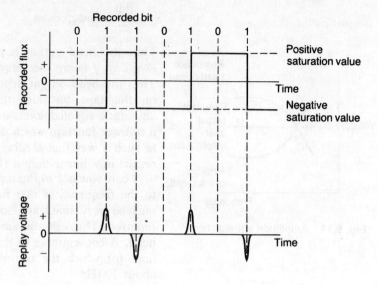

**Fig. 8.16** Non-return to zero recording

The heads are spaced across the tape and thus each lays down a track of magnetization. Thus several different signals can be simultaneously recorded.

### Example 3

What is the limiting frequency for a magnetic tape recorder which has a gap of width 5 μm and is used with a tape speed of 95.5 mm/s?

*Answer*

Using equation [11], recorded wavelength = $v/f$ with the minimum recorded wavelength as the gap width. The maximum frequency is then

$$\text{maximum frequency} = \frac{95.5 \times 10^{-3}}{5 \times 10^{-6}} = 19.1\,\text{kHz}$$

### Example 4

A magnetic tape recorder offers the option of being operated in the direct or frequency modulated modes. Which mode should be used if the signal to be recorded is (*a*) slow varying d.c. and (*b*) a high frequency 200 kHz signal?

*Answer*

(*a*) The direct mode of operation will have a bandwidth which starts at about 100 Hz and so is not suitable for slow varying d.c. The frequency modulated mode can operate down to d.c. and so is the choice.

(*b*) The frequency modulated mode has a bandwidth which is unlikely to extend beyond about 80 kHz, while the direct mode goes to much higher frequencies. The direct mode is the choice.

**Magnetic disk storage**

Digital data can be stored on magnetic disks, in a similar way to magnetic tape recorders. There may be just a single disk, generally known as a *floppy disk*, or a system of several disks on the same drive. The digital data are stored on the disk surfaces along concentric circles called *tracks*, a single disk having many such tracks. The disk is spun by the drive and the read/write heads read or write data into a track. The $3\frac{1}{2}$ inch floppy disk used in the personal computer can store 1.4 Mbytes of data.

**Problems**

1  The following are items that appear in specifications for recorders. Explain their significance.

(*a*)  Closed-loop servo recorder

Dead-band ±0.2% of span

(*b*)  UV recorder

Number of channels: 6
Chart drive: 0.16 mm/s, 0.4 mm/s, 1.6 mm/s, 4 mm/s, 10 mm/s, 25 mm/s, 100 mm/s, 250 mm/s

(*c*)  Electronic servo recorder

Maximum response time: 2 s
Dead-band ±0.3% of span maximum

(*d*)  Magnetic tape recorder

With direct record/reproduce:

**Table 8.2**

| Tape speed mm/s | Signal band Hz |
| --- | --- |
| 1524 | 300–300 000 |
| 762 | 200–150 000 |
| 381 | 100–75 000 |
| 190.5 | 100–37 500 |
| 95.5 | 100–18 750 |
| 47.6 | 100–9 300 |
| 23.8 | 100–4 650 |

(*e*)  Magnetic tape recorder

Tape speed accuracy ±0.05%

2  Compare the characteristics of the (*a*) galvanometric pen type, (*b*) knife-edge and (*c*) ultraviolet forms of recorders.

3  A galvanometer in a recorder has a resistance of 80 Ω and a damping factor of 0.7 when the transducer has a resistance of 200 Ω. How are the damping and steady state sensitivity of a recorder galvanometer affected by a resistance of 100 Ω being included in series with the transducer?

4   The coil of a UV galvanometric recorder has a resistance of $60\,\Omega$ and a damping factor of 2.8 when there is no external circuit connected to it. What additional resistance should be added to give the optimum damping factor of 0.7 when a transducer system of resistance $120\,\Omega$ is connected to the recorder?

5   A moving coil galvanometer movement has a coil of 100 turns, a cross-sectional area of $1.0 \times 10^{-4}\,\text{m}^2$, a moment of inertia of $2.4 \times 10^{-5}\,\text{kg}\,\text{m}^2$, a resistance of $60\,\Omega$ and a magnetic field of flux density 80 T. The restoring spring applies a torque of $1.2 \times 10^{-2}\,\text{N}\,\text{m}$ per radian of coil rotation. What will be (a) the steady-state voltage sensitivity, (b) the damping ratio, and (c) the natural frequency, when the signal source connected to the movement has a resistance of $240\,\Omega$?

6   Compare the performances of galvanometric recorders and potentiometric recorders.

7   What is the significance of the gap width in determining the upper frequency limit of use of a magnetic tape recorder?

8   What are the advantages and disadvantages of using a magnetic tape recorder in frequency modulated mode as compared with the direct mode?

# 9 Component measurement

## Introduction

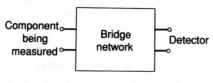

**Fig. 9.1**  Bridges

This chapter is a consideration of the methods that can be used for the measurement of resistance, capacitance, inductance, impedance and Q-factors. *Bridge* methods are widely used, bridges being essentially a network (Fig. 9.1) with two terminals across which the component being measured is connected and two terminals across which a detector is used to measure the output from the network. The output can be used as a measure of the unknown value. Often the network of the bridge is adjusted so that the detector reading becomes zero; the bridge is then said to be *balanced*. The unknown value can be determined from the values of the components in the bridge at this balance condition.

The methods used for the measurement of resistance are generally the ammeter–voltmeter method, the ohmmeter (see Chapter 6) and the Wheatstone bridge. The ammeter–voltmeter method and the ohmmeter are widely used to obtain approximate values since they are quick and simple to use. When accurate values are required, the Wheatstone bridge is used. Where very low or high resistance values have to be determined, special forms of the Wheatstone bridge can be used. The methods generally used for the measurement of capacitance, inductance, impedance and Q-factor are a.c. bridges, transformer bridges and the Q-meter.

## Ammeter–voltmeter method for resistance

The ammeter–voltmeter method for the measurement of resistance is quick and simple to use. An ammeter is connected in series with the resistance and a voltmeter in parallel with it (Fig. 9.2(*a*)), the resistance then being the ratio of their readings, i.e. *V/I*. The accuracy of this method depends on the accuracy of the meters used and the loading effect of the voltmeter (see Chapter 3). The ammeter measures not only the current through the resistance but also the current through the voltmeter. The higher the resistance of

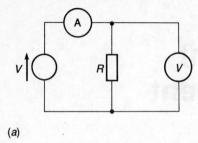

(a)

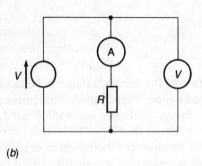

(b)

**Fig. 9.2** Ammeter–voltmeter method

the voltmeter relative to that of the resistance being measured, the smaller the fraction of the current that passes through it and so the closer the reading of the ammeter is to the current through the resistance. Loading errors are thus minimized, if high resistance voltmeters are used for the measurement of low resistances. An alternative circuit which can be used is for the voltmeter to be connected across both the ammeter and the resistance (Fig. 9.2(b)). The ammeter then measures the current through the resistance but the voltmeter reading will depart from the true value. If the unknown resistance is large compared with the resistance of the ammeter then the voltage reading is close to the true value. This circuit is thus better for the measurement of high resistances.

### Example 1

Neglecting any loading effects, what is the accuracy with which a resistance can be measured by means of the ammeter–voltmeter method if the ammeter and voltmeter have accuracies of ±5%?

*Answer*

See Chapter 1 for a discussion of the summation of errors. The percentage error in the resistance will be the sum of the percentage errors in the current and voltage, i.e. ±10%. This, however, represents the worst possible error; a more realistic value for the error is (see Chapter 1, equation [12])

$$\frac{\Delta X}{X} = \pm \sqrt{\left[\left(\frac{\Delta A}{A}\right)^2 + \left(\frac{\Delta B}{B}\right)^2\right]}$$

% error $= \pm \sqrt{[5^2 + 5^2]} = \pm 7\%$

### Balanced Wheatstone bridge

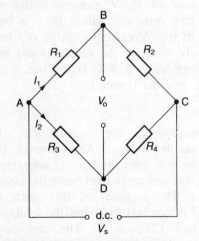

**Fig. 9.3** The Wheatstone bridge

The basic form of the *Wheatstone bridge* has a d.c. supply and each of the four bridge arms is a resistance, as shown in Fig. 9.3. The resistances in the arms of the bridge, i.e., $R_1$, $R_2$, $R_3$ and $R_4$, are so adjusted that the output potential difference $V_o$ is zero. If a galvanometer is connected between the output terminals this means the resistances are adjusted to give zero current through it. In such a condition the bridge is said to be balanced.

When the output potential difference is zero then the potential at B must equal that at D. This means that the potential difference across $R_1$, i.e. $V_{AB}$, must equal that across $R_3$, i.e., $V_{AD}$. Thus

$$I_1 R_1 = I_2 R_3$$

It also means that the potential difference across $R_2$, i.e., $V_{BC}$, must equal that across $R_4$, i.e., $V_{DC}$. Since there is no current through BD then the current through $R_2$ must be $I_1$ and that through $R_4$ must be $I_2$. Thus

$$I_1 R_2 = I_2 R_4$$

Hence

$$I_1 R_1 = I_2 R_3 = (I_1 R_2 / R_4) R_3$$

$$\frac{R_1}{R_2} = \frac{R_3}{R_4} \qquad [1]$$

The balance condition is independent of the supply voltage, depending only on the resistances in the four bridge arms. If $R_2$ and $R_4$ are known fixed resistances and $R_1$ is the unknown resistance then $R_3$ can be adjusted to give the zero potential difference condition and $R_1$ can be determined from a knowledge of the values of $R_2$, $R_3$ and $R_4$. By a suitable choice of the ratio $R_2/R_4$ a small resistance change in $R_1$ can be determined by means of a much larger resistance change in $R_3$.

The Wheatstone bridge is used for precision measurements of resistances from about $1\,\Omega$ to $1\,M\Omega$. The accuracy is mainly determined by the accuracy of the known resistors used in the bridge and the sensitivity of the null detector. Errors can also arise from changes in resistance of the bridge arms due to changes in temperature and thermoelectric e.m.f.s produced as a result of dissimilar metals being in contact. When low resistances are being measured, the resistance of the leads and contacts by which the resistor is connected to the bridge can play a significant role (see later in this chapter for a discussion of the Kelvin double bridge method of overcoming this point).

**Example 2**

A Wheatstone bridge has a resistance ratio of 1/100 for $R_2/R_4$ and $R_3$ is adjusted to give zero current. Initially this occurs with $R_3$ $1000.3\,\Omega$. The resistance $R_1$ then changes as a result of a temperature change and zero current is obtained when $R_3$ is $1002.1\,\Omega$. What was the change in resistance of $R_1$?

*Answer*

Initially

$$R_1 = \frac{R_2 R_3}{R_4} = \frac{1 \times 1000.3}{100}$$

After the change

$$R_1 + \text{change in } R_1 = \frac{1 \times 1002.1}{100}$$

$$\text{change in } R_1 = \frac{1 \times (1002.1 - 1000.3)}{100} = 0.018\,\Omega$$

Thus a change of $0.018\,\Omega$ in $R_1$ is determined by a change in $R_3$ of $1.8\,\Omega$.

**Wheatstone bridge: output voltage**

Consider the Wheatstone bridge shown in Fig. 9.3 with no galvanometer connected across the output terminals, i.e., the output load has infinite resistance. The supply voltage is connected between points A and C and thus the potential drop across the resistor $R_1$ is the fraction $R_1/(R_1 + R_2)$ of the supply voltage $V_s$. Hence

$$V_{AB} = \frac{V_s R_1}{R_1 + R_2}$$

Similarly, the potential difference across $R_3$ is

$$V_{AD} = \frac{V_s R_3}{R_3 + R_4}$$

Thus the difference in potential between points B and D, i.e., the output potential difference $V_o$, is

$$V_o = V_{AB} - V_{AD} = V_s\left(\frac{R_1}{R_1 + R_2} - \frac{R_3}{R_3 + R_4}\right) \qquad [2]$$

If $R_1$ is the unknown resistance then the relationship between its value and the output potential difference $V_o$ is a non-linear relationship. When $V_o$ is zero this equation becomes identical with the balance equation [1].

A change in resistance from $R_1$ to $R_1 + \delta R_1$ gives a change in output from $V_o$ to $V_o + \delta V_o$, hence

$$V_o + \delta V_o = V_s\left(\frac{R_1 + \delta R_1}{R_1 + \delta R_1 + R_2} - \frac{R_3}{R_3 + R_4}\right)$$

Since $V_o$ before the resistance change is given by equation [2] then

$$(V_o + \delta V_o) - V_o = V_s\left(\frac{R_1 + \delta R_1}{R_1 + \delta R_1 + R_2} - \frac{R_1}{R_1 + R_2}\right)$$

If $\delta R_1$ is much smaller than $R_1$, which is frequently the case, then the equation approximates to

$$\delta V_o = \frac{V_s \delta R_1}{R_1 + R_2} \qquad [3]$$

Under such conditions the change in output potential difference $\delta V_o$ is proportional to the change in resistance $\delta R_1$. Self-heating limits the size of the supply voltage $V_s$ and hence the size of the change in output voltage.

With the bridge the output voltage is the small difference between two larger voltages, i.e., those at B and D. An amplifier can be used to amplify this difference in voltage and ideally the amplification should be proportional to the difference in voltage and not depend on the values of the two voltages. The amplifier is said to require a high common mode

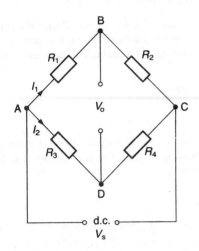

**Fig. 9.4** Thévenin equivalent circuit

rejection ratio (CMRR), the common mode signal being the average value of the two voltages at B and D. For this reason differential amplifiers are used.

The above analysis is concerned with the open circuit voltage between B and D and would only be the case if a very high resistance voltmeter was used. If, however, there is a galvanometer of resistance $R_G$ between the two points then a current $I_G$ is drawn. This current can be determined with the Thévenin equivalent circuit (Fig. 9.4). The Thévenin voltage $V_{Th}$ is the open circuit voltage $V_o$ derived above (equation [2]). Thus

$$V_{Th} = V_s \left( \frac{R_1}{R_1 + R_2} - \frac{R_3}{R_3 + R_4} \right) \qquad [4]$$

The Thévenin resistance $R_{Th}$ is the resistance seen at points B and D of the bridge and is, if the d.c. supply has negligible internal resistance,

$$R_{Th} = \frac{R_1 R_2}{R_1 + R_2} + \frac{R_3 R_4}{R_3 + R_4} \qquad [5]$$

The current $I_G$ is thus

$$I_G = \frac{V_{Th}}{R_{Th} + R_G} \qquad [6]$$

The potential difference across the galvanometer $V_G$ is

$$V_G = I_G R_G = \frac{V_{th} R_G}{R_{Th} + R_G} \qquad [7]$$

**Example 3**

A platinum resistance thermometer has a resistance at 0°C of 100 Ω and forms one arm of a Wheatstone bridge. At this temperature the bridge is balanced with each of the other arms also being 100 Ω. The temperature coefficient of resistance of the platinum is $0.0039\,°C^{-1}$. What will be the output voltage per °C change in temperature if the instrument used to measure it can be assumed to have infinite resistance and the supply voltage, with negligible internal resistance, for the bridge is 6.0 V? The resistance variation of the platinum resistance thermometer can be represented by $R_t = R_0(1 + \alpha t)$, where $R_t$ is the resistance at temperature $t$, $R_0$ it at 0°C and $\alpha$ the temperature coefficient of resistance.

*Answer*

The change in resistance of the resistance element when the temperature changes is

   change in resistance $= R_t - R_0 = R_0 \alpha t$

The change in resistance for a temperature change of 1°C is thus

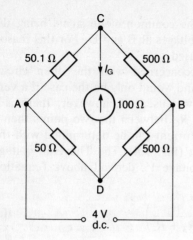

**Fig. 9.5** Example 4

change in resistance = $100 \times 0.0039 \times 1 = 0.39\,\Omega$

Since this resistance is small compared with the $100\,\Omega$ and the load across the output terminals is effectively infinite then the approximate equation [3], derived for the open circuit voltage, can be used. Thus

$$\delta V_\mathrm{o} = \frac{V_\mathrm{s}\delta R_1}{R_1 + R_2} = \frac{6.0 \times 0.39}{100 + 100} = 0.012\,\mathrm{V}$$

## Example 4

For the Wheatstone bridge shown in Fig. 9.5 what will be the out-of-balance current through the galvanometer? The d.c. supply may be assumed to have negligible resistance.

*Answer*

The Thévenin equivalent circuit for the bridge is shown in Fig. 9.6. The Thévenin equivalent resistance is given by equation [5] as

$$R_\mathrm{Th} = \left(\frac{500 \times 50.1}{500 + 50.1} + \frac{500 \times 50}{500 + 50}\right) = 90.99\,\Omega$$

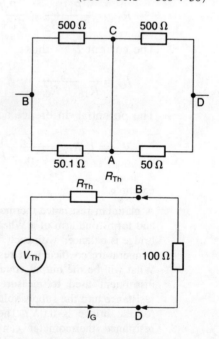

**Fig. 9.6** Example 4

The Thévenin voltage is given by equation [4] as

$$V_\mathrm{Th} = 4\left(\frac{50.1}{50.1 + 500} - \frac{50}{50 + 500}\right) = 6.61 \times 10^{-4}\,\mathrm{V}$$

Hence the current $I_\mathrm{G}$ through the galvanometer is given by equation [6] as

$$I_\mathrm{G} = \frac{V_\mathrm{Th}}{R_\mathrm{Th} + R_\mathrm{G}} = \frac{6.61 \times 10^{-4}}{90.99 + 100} = 3.46 \times 10^{-6}\,\mathrm{A}$$

**Wheatstone bridge: compensation**

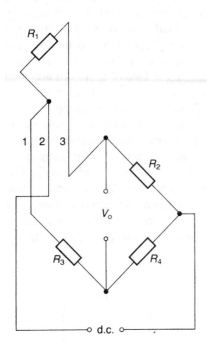

**Fig. 9.7** Compensation for leads

In many measurement systems involving a resistive transducer the actual resistance element may have to be at the end of long leads. The resistance of these leads will be affected by changes in temperature. For example, a platinum resistance thermometer consists of a platinum coil at the ends of leads. When the temperature changes not only will the resistance of the platinum coil change but so also will the resistance of the leads. What is required is just the resistance of the coil and so some means has to be employed to compensate for the lead resistance. One method of doing this is to use three leads to the coil, as shown in Fig. 9.7. The coil is then connected into a Wheatstone bridge in such a way that lead 1 is in series with the $R_3$ resistor while lead 3 is in series with the platinum resistance coil $R_1$. Lead 2 is the connection to the power supply. Any change in lead resistance as a result of a temperature change will affect all three leads equally since they are all the same length and resistance. The result is that changes in lead resistance occur equally in two arms of the bridge and will cancel out if $R_1$ and $R_3$ are the same resistance.

The electrical resistance strain gauge is another transducer where compensation has to be made for temperature effects. The strain gauge changes resistance as the strain applied to it changes. Unfortunately it also changes resistance if the temperature changes. One way of eliminating the temperature effect is to use a dummy gauge. This is a strain gauge which is identical to the one under strain, the active gauge, but is not subject to the strain. It is, however, at the same temperature as the active gauge. Thus a temperature change will cause both gauges to change resistance by the same amount. The active gauge is mounted in one arm of a Wheatstone bridge and the dummy gauge in another arm such that the effects of temperature-induced resistance changes cancel out (Fig. 9.8).

Strain gauges are often used as a secondary transducer, i.e., a means of converting the output of the initial transducer into

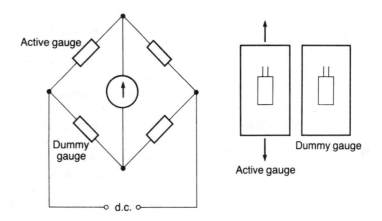

**Fig. 9.8** Compensation with strain gauges

a more convenient form. For example, a load cell suffers elastic deformation as the result of forces and this elastic deformation can be converted into a resistance change by strain gauges attached to the load cell. In such a situation temperature effects on the strain gauges have to be compensated for. While dummy gauges could be used, a better solution is to use four strain gauges. Two of them are attached to the load cell in such a way that they are in tension and two so that they are in compression when the forces are applied (Fig. 9.9). The gauges that are in tension will increase in resistance while those in compression will decrease in resistance as a result of the application of the forces. If the gauges are connected as the four arms of a Wheatstone bridge, as in Fig. 9.9, then since all will be equally affected by any temperature changes the arrangement is temperature-compensated. The arrangement also gives a much greater output voltage or galvanometer current than using just a single active gauge.

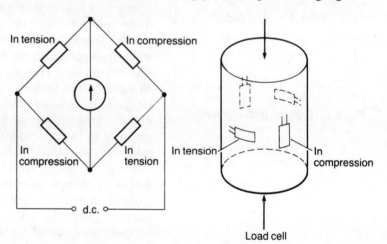

**Fig. 9.9**   Four active arm strain gauge bridge

### Kelvin double bridge

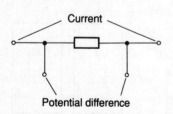

**Fig. 9.10**   Four-terminal resistor

A modification of the Wheatstone bridge that can be used for the measurement of low values of resistance, below about $1\,\Omega$, is the *Kelvin double bridge*. With such low resistances it is necessary to obtain an accurate definition of the resistance being measured by using *four-terminal resistors* (Fig. 9.10). Two of the terminals define the points between which the current is supplied and two the points between which the potential difference is determined. Figure 9.11 shows the form of the Kelvin double bridge. $R_1$ is the resistance being measured and $R_2$ a standard resistance of about the same size. The resistance of the connector linking the two resistors $R_1$ and $R_2$, sometimes referred to as the yoke, is $r$. The resistances $R_3$, $R_4$, $r_3$ and $r_4$ have either $R_3$ and $r_3$ or $R_4$ and $r_4$ variable with the relationship between their resistances fixed as

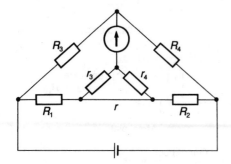

**Fig. 9.11**   Kelvin double bridge

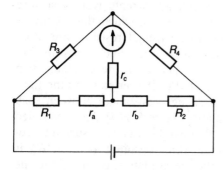

**Fig. 9.12**   Equivalent circuit

$$\frac{R_3}{R_4} = \frac{r_3}{r_4} \qquad [8]$$

Using the delta star transformation, the delta part of the bridge can be transformed into a star to give the equivalent circuit shown in Fig. 9.12, with

$$r_a = \frac{r_3 r}{r_3 + r_4 + r}$$

$$r_c = \frac{r_4 r}{r_3 + r_4 + r}$$

The balance condition for the bridge is the same as that for the Wheatstone bridge and is thus

$$\frac{R_1 + r_a}{R_2 + r_c} = \frac{R_3}{R_4}$$

Hence, rearranging this equation gives

$$R_1 = \frac{R_3(R_2 + r_c)}{R_4} - r_a$$

Substituting for $r_a$ and $r_c$ gives

$$R_1 = \frac{R_3 R_2}{R_4} + \frac{R_3 r_4 r}{r_3 + r_4 + r} - \frac{r_3 r}{r_3 + r_4 + r}$$

$$= \frac{R_3 R_2}{R_4} + \frac{r_4 r}{r_3 + r_4 + r}\left(\frac{R_3}{R_4} - \frac{r_3}{r_4}\right) \qquad [9]$$

But $R_3/R_4$ equals $r_3/r_4$ (equation [8]), hence the balance condition with this condition becomes simplified to

$$R_1 = \frac{R_3 R_2}{R_4} \qquad [10]$$

**Example 5**

A Kelvin double bridge, of the form shown in Fig. 9.11, is used with $R_3/R_4 = r_3/r_4$. At balance, $R_4 = 100\,\Omega$, $R_2 = 9.7\,\Omega$ and $R_3 = 0.1\,\Omega$. What is the value of the resistance $R_1$?

*Answer*

Because $R_3/R_4 = r_3/r_4$, equation [10] can be used for the balance condition. Thus

$$R_1 = \frac{R_3 R_2}{R_4} = \frac{0.1 \times 9.7}{100} = 0.0097\,\Omega$$

**High resistance bridge**

The conventional form of Wheatstone bridge is not capable of being used for the measurement of very high resistances, i.e.,

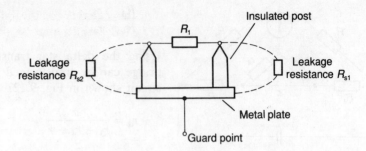

**Fig. 9.13**  Three-terminal resistor

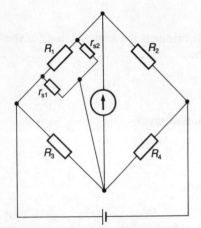

**Fig. 9.14**  Guarded bridge

**a.c. bridge**

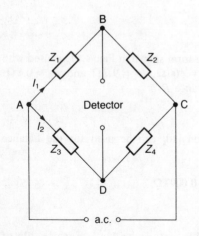

**Fig. 9.15**  Basic a.c. bridge

resistances of the order of thousands of megaohms. A major problem that occurs is that of leakage currents over the surface of, for example, the insulated posts on which the resistor is mounted. For this reason, three-terminal resistors are often used; Fig. 9.13 shows the basic form. Such resistors are mounted on insulating pillars above a metal plate, two terminals being directly connected to the resistor and the third to the plate. There will be leakage resistances between each resistor terminal and the plate. Figure 9.14 shows a modified form of Wheatstone bridge that can be used with such a three-terminal resistor. When the three-terminal resistor $R_1$ is connected into the bridge, the metal plate is connected to the detector junction with $R_3$ and $R_4$. This puts the leakage resistance $R_{s1}$ in parallel with $R_3$ and since $R_{s1}$ is much larger than $R_3$ its effect is negligible. The leakage resistance $R_{s2}$ is in parallel with the detector and thus its only effect is to change the detector sensitivity.

When the resistors in the arms of a Wheatstone bridge are replaced by impedances and an a.c. source used instead of a d.c. source, the bridge is called an *a.c. bridge* (Fig. 9.15). Impedance is the quantity obtained by dividing the potential difference phasor by the current phasor. Since both these phasors have magnitudes and phases, impedance has also a phase $\phi$ and magnitude $|Z|$ and can be written as $|Z|\angle\phi$. Thus for zero potential difference between A and D, and hence zero current through the detector, the potential difference across $Z_1$ must be the same as that across $Z_3$, in both magnitude and phase, and similarly the potential differences across $Z_2$ and $Z_4$ must be equal in both magnitude and phase. Thus

$$I_1|Z_1|\angle\phi_1 = I_2|Z_3|\angle\phi_3$$

$$I_1|Z_2|\angle\phi_2 = I_2|Z_4|\angle\phi_4$$

where the currents $I_1$ and $I_2$ are phasors. Hence, dividing the two equations gives

$$\frac{|Z_1|\angle\phi_1}{|Z_2|\angle\phi_2} = \frac{|Z_3|\angle\phi_3}{|Z_4|\angle\phi_4} \qquad [11]$$

For this equality to be valid, the magnitudes must balance and the phase angles must balance. For the magnitudes to balance

$$\frac{|Z_1|}{|Z_2|} = \frac{|Z_3|}{|Z_4|}$$ [12]

This can be written as $|Z_1||Z_4| = |Z_2||Z_3|$, i.e., the *products of the magnitudes of the opposite arms must be equal*.

For the phases to balance

$$\angle \phi_1 - \angle \phi_2 = \angle \phi_3 - \angle \phi_4$$

This can be written as

$$\angle \phi_1 + \angle \phi_4 = \angle \phi_2 + \angle \phi_3$$ [13]

i.e., the *sum of the phase angles of the opposite arms must be equal*.

The above representation of the impedance is called polar notation. An alternative notation is the representation of impedance as the sum of a real term and a complex term, i.e., $R + jX$ where $R$ is the resistance and $X$ the reactance. The magnitude of the impedance is then $\sqrt{(R^2 + X^2)}$ and the phase $\phi$ is $\tan^{-1} X/R$. With this form of representation equation [11] becomes

$$\frac{R_1 + jX_1}{R_2 + jX_2} = \frac{R_3 + jX_3}{R_4 + jX_4}$$ [14]

This equation can be rearranged to give the sum of a real number and an imaginary number on each side of the equals sign. For two complex numbers to be equal both their real terms and their imaginary terms must be equal. This thus becomes the balance conditions when complex notation is used.

There are many variations of the basic a.c. bridge, the following are some of the more common ones and the interpretation of the above conditions for balance for each bridge.

### Example 6

For the a.c. bridge shown in Fig. 9.15 the impedances are $Z_2 = 100 \angle 80° \Omega$, $Z_3 = 200 \Omega$, and $Z_4 = 400 \angle 30° \Omega$. What is the value of $Z_1$ at balance?

*Answer*

The products of the magnitudes of the opposite arms must be equal. Thus

$$Z_1 \times 400 = 100 \times 200$$

Hence the magnitude is $50\,\Omega$. The sum of the phase angles of the opposite arms must also be equal. Thus

$$\angle\phi_1 + 30° = 80° + 0°$$

Hence the phase of $Z_1$ is $50°$. Thus $Z_1$ is $50\angle 50°\,\Omega$.

**Equivalent circuits for R, L and C**

Resistors, inductors and capacitors do not exist as pure components with resistors only having resistance, inductors only inductance and capacitors only capacitance. In general, they each have resistance, inductance and capacitance. At its simplest, an inductor can be considered to be a pure inductance in series with, or sometimes in parallel with, a pure resistance, this being a resistance term resulting from the resistance of the windings of the inductor, eddy current losses and hysteresis losses if the inductor is iron-cored. With a capacitor, the simplest equivalent is a pure capacitance shunted by, or in series with, a pure resistance. The resistance term results from leakage resistance and dielectric losses.

The Q-factor is used as an expression of the quality of an inductor or capacitor, though for a capacitor the *dissipation factor D*, which is $1/Q$, is more widely used. The Q-factor is defined as·

$$Q = \frac{2\pi \times \text{maximum energy stored in a cycle}}{\text{energy dissipated per cycle}} \qquad [15]$$

and is a measure of the ability of the component to store and then release energy. The Q-factor can be shown to be, for a capacitor considered as capacitance in series with resistance or an inductor as inductance in series with resistance,

$$Q = \frac{\text{reactance}}{\text{resistance}} \qquad [16]$$

A high Q-factor for an inductor would thus indicate that the series resistance term is small compared with the reactance due to the inductance. A low dissipation factor for a capacitor would indicate a high Q-factor and that the series resistance term is small compared with the reactance due to the capacitance. Typical dissipation factors for commercial capacitors range from about 0.0001 to 0.001.

**Maxwell–Wien bridge**

The Maxwell–Wien bridge (Fig. 9.16) is used for the determination of the inductance and resistance of inductors and is most suited to those having medium Q-factors, i.e., between about 1 and 10. To write the balance equation, the first step is to write each impedance in complex notion. Thus, since $Z_1$ consists of a resistor in parallel with a capacitor

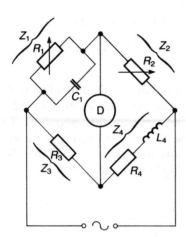

**Fig. 9.16** Maxwell–Wien bridge

$$\frac{1}{Z_1} = \frac{1}{R_1} + j\omega C_1$$

$$Z_1 = \frac{R_1}{1 + j\omega C_1 R_1}$$

$Z_2$ and $Z_3$ are purely resistances with $Z_2 = R_2$ and $Z_3 = R_3$. $Z_4$ is an inductor having both resistance and inductance, these being the quantities to be determined. Thus

$$Z_4 = R_4 + j\omega L_4$$

Hence for the impedances to balance we must have

$$\frac{Z_1}{Z_2} = \frac{Z_3}{Z_4}$$

$$Z_4 = \frac{Z_2 Z_3}{Z_1}$$

$$R_4 + j\omega L_4 = \frac{R_2 R_3 (1 + j\omega C_1 R_1)}{R_1}$$

For the real parts we have a balance condition of

$$R_4 = \frac{R_2 R_3}{R_1} \qquad\qquad [17]$$

For the imaginary parts

$$L_4 = R_2 R_3 C_1 \qquad\qquad [18]$$

The balance conditions are independent of the frequency of the bridge a.c. supply. The bridge is widely used for the determination of the series resistance $R_4$ and the inductance $L_4$ of an inductor. The procedure is usually to adjust $R_2$ to obtain the best balance, then $R_1$ to improve it and then $R_2$ again and so on until a final balance is obtained. The Q-factor for the inductor is

$$Q = \frac{\omega L_4}{R_4} = \frac{\omega R_2 R_3 C_1}{R_2 R_3 / R_1} = \omega R_1 C_1 \qquad\qquad [19]$$

### Example 7

A Maxwell bridge with a 1 kHz a.c. supply is used to determine the inductance and series resistance of an inductor. At balance, the bridge arms are AB 2.0 µF in parallel with 10 kΩ, BC 200 Ω, CD the inductor, and DA 300 Ω. What are the inductance, series resistance and Q-factor of the inductor?

*Answer*

Using equations [17], [18] and [19]

$$R_4 = \frac{R_2 R_3}{R_1} = \frac{200 \times 300}{10 \times 10^3} = 6.0\,\Omega$$

$$L_4 = R_2R_3C_1 = 200 \times 300 \times 2.0 \times 10^{-6} = 0.12\,\text{H}$$

$$Q = \omega C_1 R_1 = 1000 \times 2.0 \times 10^{-6} \times 10 \times 10^3 = 20$$

## The Hay bridge

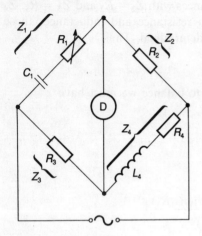

**Fig. 9.17** Hay bridge

The Hay bridge (Fig. 9.17) is used for the determination of the inductance and resistance of inductors and is most suited to those having high Q-factors, i.e., greater than about 10. The bridge differs from the Maxwell bridge in having a variable resistor in series with the capacitor instead of in parallel. Writing each impedance in complex notation gives

$$Z_1 = R_1 + \frac{1}{j\omega C_1} = R_1 - \frac{j}{\omega C_1}$$

The pure resistances $Z_2 = R_2$ and $Z_3 = R_3$.

$$Z_4 = R_4 + j\omega L_4$$

Hence for the impedances to balance we must have

$$\frac{Z_1}{Z_2} = \frac{Z_3}{Z_4}$$

$$Z_4 = \frac{Z_2 Z_3}{Z_1}$$

$$R_4 + j\omega L_4 = \frac{R_2 R_3}{R_1 - (j/\omega C_1)}$$

$$R_4 R_1 + \frac{L_4}{C_1} + j\omega L_4 R_1 - \frac{jR_4}{\omega C_1} = R_2 R_3$$

Equating the real terms gives

$$R_4 R_1 + \frac{L_4}{C_1} = R_2 R_3 \qquad [20]$$

and equating the imaginary terms gives

$$\omega L_4 R_1 = \frac{R_4}{\omega C_1} \qquad [21]$$

Using equation [21] to substitute for $L_4$ in equation [20],

$$R_4 R_1 + \frac{R_4}{\omega^2 R_1 C_1^2} = R_2 R_3$$

$$R_4 = \frac{\omega^2 R_1 R_2 R_3 C_1^2}{1 + \omega^2 R_1^2 C_1^2} \qquad [22]$$

Substituting this value of $R_4$ in equation [21] gives

$$L_4 = \frac{R_2 R_3 C_1}{1 + \omega^2 R_1^2 C_1^2} \qquad [23]$$

The Q-factor is

$$Q = \frac{\omega L_4}{R_4} = \frac{\omega R_2 R_3}{\omega^2 R_1 R_2 R_3 C_1^2} = \frac{1}{\omega R_1 C_1} \qquad [24]$$

Equations [22] and [23] can thus be used to obtain values for $R_4$ and $L_4$. However, they contain the frequency and so it is necessary to have an accurate value of that quantity. However, if the bridge is used for a high $Q$ inductor the expressions can be simplified.

Equation [23] can be written, using equation [24], as

$$L_4 = \frac{R_2 R_3 C_1}{1 + (1/Q^2)} \qquad [25]$$

When $Q$ is greater than about 10 the $(1/Q^2)$ term becomes insignificant and so

$$L_4 = R_2 R_3 C_1 \qquad [26]$$

**Example 8**

A Hay bridge has an a.c. source of frequency 1 kHz and at balance the bridge arms are AB 0.1 μF in series with 95 Ω, BC 900 Ω, CD the unknown inductor, and DA 500 Ω. What are the Q-factor and inductance of the inductor?

*Answer*

The Q-factor for the inductor is given by equation [24] as

$$Q = \frac{1}{\omega C_1 R_1} = \frac{1}{2\pi \times 1000 \times 0.1 \times 10^{-6} \times 95} = 16.8$$

Thus the simplified equation for balance can be used. Hence, using equation [26]

$$L_4 = R_2 R_3 C_1 = 900 \times 500 \times 0.1 \times 10^{-6} = 45 \, \text{mH}$$

**The Owen bridge**

The Owen bridge (Fig. 9.18) is used as a high precision bridge for the measurement of the inductance and resistance of inductors. Writing each impedance in complex form gives

$$Z_1 = \frac{1}{j\omega C_1}$$

$$Z_2 = R_2$$

$$Z_3 = R_3 + \frac{1}{j\omega C_3} = R_3 - \frac{j}{\omega C_3}$$

$$Z_4 = R_4 + j\omega L_4$$

Hence for the impedances to balance

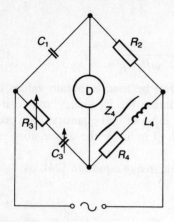

**Fig. 9.18** Owen bridge

$$\frac{Z_1}{Z_2} = \frac{Z_3}{Z_4}$$

$$Z_4 = \frac{Z_2 Z_3}{Z_1}$$

$$R_4 + j\omega L_4 = j\omega C_1 R_2 \left(R_3 - \frac{j}{\omega C_3}\right) = j\omega C_1 R_2 R_3 + \frac{C_1 R_2}{C_3}$$

Equating the real terms gives

$$R_4 = \frac{C_1 R_2}{C_3} \qquad\qquad [27]$$

Equating the imaginary terms gives

$$L_4 = C_1 R_2 R_3 \qquad\qquad [28]$$

The balance condition is independent of the frequency of the a.c. bridge supply.

**The series capacitance bridge**

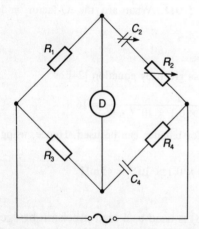

**Fig. 9.19** Series capacitance bridge

The series capacitance bridge (Fig. 9.19) is widely used for the measurement of the capacitance and series resistance of capacitors. Writing each impedance in complex form gives

$$Z_1 = R_1$$

$$Z_2 = R_2 + \frac{1}{j\omega C_2} = R_2 - \frac{j}{\omega C_2}$$

$$Z_3 = R_3$$

$$Z_4 = R_4 + \frac{1}{j\omega C_4} = R_4 - \frac{j}{\omega C_4}$$

Hence for the impedances to balance

$$\frac{Z_1}{Z_2} = \frac{Z_3}{Z_4}$$

$$Z_4 = \frac{Z_2 Z_3}{Z_1}$$

$$R_4 - \frac{j}{\omega C_4} = \frac{R_3}{R_1}\left(R_2 - \frac{j}{\omega C_2}\right)$$

Hence equating the real terms gives

$$R_4 = \frac{R_3 R_2}{R_1} \qquad\qquad [29]$$

Equating the imaginary terms gives

$$-\frac{1}{\omega C_4} = -\frac{R_3}{\omega C_2 R_1}$$

$$C_4 = \frac{C_2 R_1}{R_3} \qquad [30]$$

The dissipation factor $D$ is

$$D = \frac{R_4}{1/\omega C_4} = \omega R_4 C_4 = \omega R_2 C_2 \qquad [31]$$

**Example 9**

A series capacitance bridge with an a.c. source of 10 kHz has at balance arms of AB 7 kΩ, BC 500 pF with 160 Ω, CD the unknown capacitor and DA 10 kΩ. What are the capacitance, series resistance and dissipation factor of the unknown capacitor?

*Answer*

Using equation [30]

$$C_4 = \frac{C_2 R_1}{R_3} = \frac{500 \times 10^{-12} \times 7 \times 10^3}{10 \times 10^3} = 350 \, \text{pF}$$

Using equation [29]

$$R_4 = \frac{R_3 R_2}{R_1} = \frac{10 \times 10^3 \times 160}{7 \times 10^3} = 229 \, \Omega$$

Using equation [31]

$$D = \omega C_2 R_2 = 2\pi \times 10 \times 10^3 \times 500 \times 10^{-12} \times 160 = 5.0 \times 10^{-3}$$

**The parallel capacitance bridge**

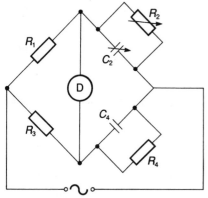

**Fig. 9.20**  Parallel capacitance bridge

The parallel capacitance bridge is particularly suitable for capacitor measurements when the dissipation factor is high, e.g., 0.05 or more. With this bridge the unknown capacitor is considered to be a pure capacitance in parallel with resistance. Figure 9.20 shows the bridge. The balance conditions for the bridge are derived in the same way as the previous bridges and are

$$C_4 = \frac{R_1 C_2}{R_3} \qquad [32]$$

$$R_4 = \frac{R_3 R_2}{R_1} \qquad [33]$$

$$D = \omega C_2 R_2 \qquad [34]$$

**The Wien bridge**

The Wien bridge (Fig. 9.21) is used for the measurement of capacitors when they are considered in terms of a pure capacitance in parallel with resistance. The bridge is also used as a frequency-dependent circuit in oscillators. The balance conditions are derived in the same way as the previous bridges and are

$$C_4 = \frac{(R_1/R_2) C_3}{1 + \omega^2 R_3^2 C_3^2} \qquad [35]$$

$$R_4 = \frac{R_2(1 + \omega^2 R_3^2 C_3^2)}{\omega^2 R_3 R_1 C_3^2} \qquad [36]$$

$$D = \omega C_3 R_3 \qquad [37]$$

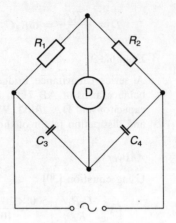

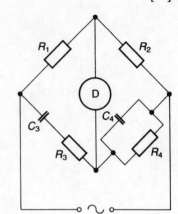

**Fig. 9.21** Wien bridge

## The Schering bridge

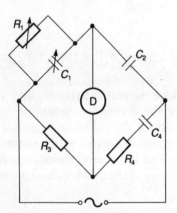

**Fig. 9.22** Schering bridge

The Schering bridge (Fig. 9.22) is used for the measurement of capacitors in terms of a pure capacitance in series with resistance and is generally used for capacitors with very low dissipation factors. The conditions for balance are derived in the same way as previous bridges and are

$$C_4 = \frac{R_1 C_2}{R_3} \qquad [38]$$

$$R_4 = \frac{R_3 C_1}{C_2} \qquad [39]$$

$$D = \omega C_1 R_1 \qquad [40]$$

### Example 10

A Schering bridge has a 10 kHz a.c. source and at balance has arm $R_1$ 1050 Ω, $C_1$ 205 pF, $C_2$ 10 pF, and $R_3$ 20 Ω. What are the capacitance, series resistance and dissipation factor of the capacitor in the fourth arm?

*Answer*

Using equation [38]

$$C_4 = \frac{R_1 C_2}{R_3} = \frac{1050 \times 10 \times 10^{-12}}{20} = 525 \, \text{pF}$$

Using equation [39]

$$R_4 = \frac{R_3 C_1}{C_2} = \frac{20 \times 205 \times 10^{-12}}{10 \times 10^{-12}} = 410 \, \Omega$$

Using equation [40]

$$D = \omega C_1 R_1 = 10 \times 10^3 \times 205 \times 10^{-12} \times 1050 = 0.0022$$

**Stray impedances with a.c. bridges**

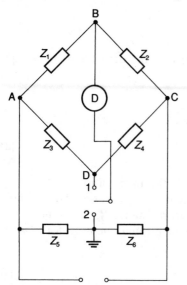

**Fig. 9.23**  Wagner earth

There will be stray capacitances between the various elements of an a.c. bridge and the ground. The effects of these capacitances can be minimized by shielding the bridge elements and earthing the shields. The stray capacitances between the detector terminals and earth can be eliminated by using a *Wagner earth* (Fig. 9.23). This is a means of ensuring that the points B and D of a balanced bridge are at ground potential. With the switch in position 1 the bridge is balanced by adjusting $Z_3$. Then with the switch in position 2 the bridge with $Z_5$ and $Z_6$ is balanced by adjusting $Z_3$ and $Z_4$. This ensures that D is at earth potential. The switch is then put back to position 1 and the balancing process repeated, then to position 2 and is again repeated. The two balancing processes are repeated until the bridge remains balanced when switched between the two points.

**Transformer ratio bridges**

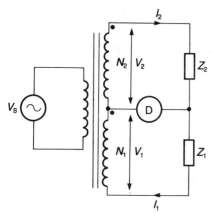

**Fig. 9.24**  Single ratio transformer bridge

The bridges described so far in this chapter attain balance by adjustment of impedances. The *transformer ratio bridge*, however, is balanced by varying the turns ratio of a transformer. Such a bridge has many advantages over a.c. bridges: only a small number of standard resistors and capacitors are required, the problem of stray impedances is largely eliminated, the turns ratio can be accurately determined and is not affected by temperature or environmental changes. Hence the transformer ratio bridge is widely used as a universal bridge for the measurement of the resistance, inductance and capacitance of components, being used up to about 250 MHz.

Figure 9.24 shows the basic form of a *single ratio transformer bridge*. A tapped transformer is used to provide a voltage division of the source voltage $V_s$ which is dependent on the number of turns across which the tapping is made. The voltage across $N_1$ turns is proportional to $N_1$, and that across $N_2$ proportional to $N_2$ and so

$$V_1 = kN_1 = I_1 Z_1$$

$$V_2 = kN_2 = I_2 Z_2$$

where $k$ is a constant. Thus the current $I_1$ through the impedance $Z_1$ is

$$I_1 = \frac{kN_1}{Z_1}$$

and the current $I_2$ through $Z_2$ is

$$I_2 = \frac{kN_2}{Z_2}$$

For there to be no current through the detector, $I_1$ must equal $I_2$, hence

$$\frac{N_1}{Z_1} = \frac{N_2}{Z_2}$$

$$\frac{Z_1}{Z_2} = \frac{N_1}{N_2} \qquad [41]$$

The ratio of the impedances is the turns ratio.

For resistance measurements $Z_1$ can be the unknown resistance $R_x$ and $Z_2$ a standard resistance $R_s$. Thus the ratio of the resistances when the current through the detector is zero is the ratio of the turns tapped:

$$R_x = \frac{N_1 R_s}{N_2} \qquad [42]$$

For capacitance measurement $Z_1$ can be the unknown capacitor with capacitance $C_x$ and parallel resistance of $R_x$, and $Z_2$ a standard capacitor $C_s$ with a parallel variable resistance $R_s$. Thus

$$\frac{1}{Z_1} = \frac{1}{R_x} + j\omega C_x = \frac{1 + j\omega C_x R_x}{R_x}$$

$$\frac{1}{Z_2} = \frac{1}{R_s} + j\omega C_s = \frac{1 + j\omega C_s R_s}{R_s}$$

Hence, at balance, equation [42] can be written as

$$\frac{1}{Z_1} = \left(\frac{N_2}{N_1}\right)\frac{1}{Z_2}$$

and so

$$\frac{1 + j\omega C_x R_x}{R_x} = \left(\frac{N_2}{N_1}\right)\left(\frac{1 + j\omega C_s R_s}{R_s}\right)$$

Hence equating the real terms

$$R_x = \frac{N_1 R_s}{N_2} \qquad [43]$$

and equating the imaginary terms

$$C_x = \frac{N_2 C_s}{N_1} \qquad [44]$$

Inductors can be considered as inductance in parallel with

resistance and the comparison made with a capacitor in parallel with a resistance. To obtain a balance it is necessary to reverse the current direction through the winding connected to the capacitance standard. Thus

$$\frac{1}{Z_1} = \frac{1}{R_x} + \frac{1}{j\omega L_x} = \frac{1}{R_x} - \frac{j}{\omega L_x}$$

$$\frac{1}{Z_2} = \frac{1}{R_s} - j\omega C_s$$

The minus sign in the above equation is because of the reversed connection. Thus at balance equation [42] can be written as

$$\frac{1}{Z_1} = \left(\frac{N_2}{N_1}\right)\frac{1}{Z_2}$$

$$\frac{1}{R_x} - \frac{j}{\omega L_x} = \left(\frac{N_2}{N_1}\right)\left(\frac{1}{R_s} - j\omega C_s\right)$$

Hence equating the imaginary terms

$$\frac{1}{\omega L_x} = \frac{N_2 \omega C_s}{N_1}$$

$$L_x = \frac{N_1}{N_2 \omega^2 C_s} \tag{45}$$

Equating the real terms

$$R_x = \frac{N_1 R_s}{N_2} \tag{46}$$

The transformer used is generally a multi-decade ratio transformer. This has sets of tappings, each of which has nine tappings with the first decade being for $n$ turns per tap, the second decade for $10n$ turns per tap, the third for $100n$ turns per tap, etc. The single ratio transformer bridge has the disadvantage that the leakage inductance and winding resistance of the transformer windings is in series with $Z_1$ and $Z_2$. For this reason the bridge is mainly used for the measurement of high impedances when the impedance due to the leakage inductance and winding resistance is negligible.

The range over which measurements can be made can be extended by using a double ratio transformer bridge (Fig. 9.25). When the detector indicates zero current then there is zero magnetic flux in the core of the second transformer. For this to occur the ampere turns $n_1 I_1$ and $n_2 I_2$ must be equal and producing fluxes which oppose each other. But $I_1 = V_1/Z_1$ with $V_1 = kN_1$, and $I_2 = V_2/Z_2$ with $V_2 = kN_2$. Hence

$$\frac{n_1 N_1}{Z_1} = \frac{n_2 N_2}{Z_2} \tag{47}$$

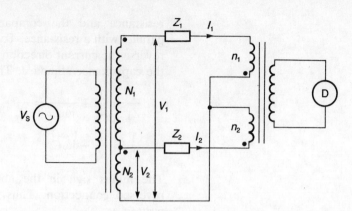

**Fig. 9.25** Double ratio transformer bridge

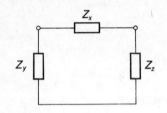

**Fig. 9.26** $Z_x$ in a circuit

Double ratio transformer bridges can be used to measure the impedances of components *in situ*, i.e., while part of some other circuit. A conventional bridge used to measure the impedance $Z_x$ in Fig. 9.26 would also give a result which was the impedance of the parallel arrangement with $Z_y$ and $Z_z$. If, however, a double ratio bridge is used, as shown in Fig. 9.27, then though $Z_y$ shunts the input transformer it will generally have negligible effect on the output from those turns. $Z_z$ will have no effect at balance because there is then no voltage across it. The measurement thus becomes that of $Z_x$.

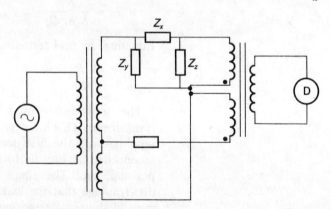

**Fig. 9.27** Measurement *in situ*

A *universal bridge* has standard capacitors and resistances which can be switched across different tappings on the transformer and so enable different ranges to be obtained. Typically such a bridge is used over a frequency range of about 200 Hz to 10 kHz, measuring capacitances in the range 0.1 pF to 10 mF, inductors 0.1 μH to 10 kH and resistance 0.001 Ω to 1 GΩ with an accuracy of the order of ±0.1%.

**Example 11**

A single ratio transformer bridge with $N_1 = N_2$ is used to measure the parallel components of an unknown impedance at a frequency of

COMPONENT MEASUREMENT 137

10 kHz. The bridge was balanced using standard capacitors and resistors and the scaled results obtained were − 2.50 nF and 180 Ω. What are the parallel components?

*Answer*

The negative sign indicates that the component being measured is an inductor. The balance conditions are thus given by equations [45] and [46]:

$$L_x = \frac{N_1}{N_2 \omega^2 C_s} = \frac{1}{(2\pi \times 10^4)^2 \times 2.50 \times 10^{-9}} = 0.10\,\mathrm{H}$$

$$R_x = \frac{N_1 R_s}{N_2} = 180\,\Omega$$

**Q-meter**

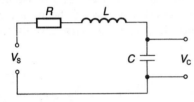

**Fig. 9.28**   Series resonant circuit

For a series *RLC* circuit (Fig. 9.28) at resonance the Q-factor is the reactance divided by the resistance and so is

$$Q = \frac{1}{\omega_0 CR} \qquad [48]$$

where $\omega_0$ is the resonant frequency. At resonance the current $I_0$ in the circuit is determined only by the circuit resistance $R$ and is thus $V_s/R$. Therefore at the resonant frequency the voltage $V_C$ across the capacitor is $I_0 X_C$ and so

$$V_C = \frac{V_s}{R} \times \frac{1}{\omega_0 C} = V_s Q \qquad [49]$$

The voltage across the capacitor is thus the Q-factor multiplied by $V_s$. Measurement of this voltage is the basis of the Q-meter.

Figure 9.29 shows the basic Q-meter circuit. An oscillator passes current through a very low resistance, of the order of 0.02 Ω, which is in the resonant circuit. The potential difference across this resistor acts as a voltage source $V_s$ with a very small internal resistance. This source is in a series *RLC* circuit which can be tuned to the resonant frequency by means of a variable capacitor, the circuit in Fig. 9.29 being shown in a form suitable for the measurement of an inductor. The voltage across the variable capacitor $V_C$ is measured by an electronic voltmeter which has a scale directly giving Q-factor values. However, $V_C$ is only the Q-value if $V_s$ is 1. Thus the values given have to be multiplied by the factor indicated by the meter used to measure $V_s$. This is usually a thermocouple meter and it has a scale giving the factor by which the $V_C$ value must be multiplied. The inductance of the unknown inductor can be determined from the fact that at the resonant condition the reactance of the inductor equals the reactance of the capacitor. Thus

$$\omega_0 L = \frac{1}{\omega_0 C} \qquad [50]$$

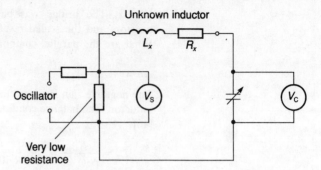

Unknown inductor

**Fig. 9.29**   Q-meter circuit

Hence, knowing the value of the tuning capacitor and the resonant frequency, the inductance can be calculated. This assumes that the only significant inductance in the circuit is that of the unknown inductor. This is a reasonable approximation if the inductance being measured is large. The resistance of the inductor can be determined in terms of the indicated Q-factor of the circuit, since it constitutes virtually all the resistance in the series circuit:

$$\omega_0 L = \frac{1}{\omega_0 C} = QR_x \qquad\qquad [51]$$

Low impedances, such as large capacitances, small inductances and low resistances, are determined by connecting the unknown component such as a capacitor $C_x$ in series with the variable capacitor and an inductor (Fig. 9.30). The unknown capacitor is first short-circuited and the circuit tuned to give a Q-factor value. If the variable capacitor for this has a value $C_1$ and the frequency is $\omega_0$ then

$$Q_1 = \frac{1}{\omega_0 C_1 R}$$

Then the short-circuit is removed and the circuit tuned to the same frequency as before by adjusting the variable capacitor

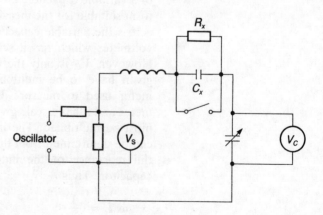

**Fig. 9.30**   Measurement of a low impedance

to a value $C_2$. The new Q-factor is $Q_2$. The capacitance in the circuit is $C_2$ in series with $C_x$. Since the resonant frequency is the same as when $C_x$ is short-circuited, this capacitance must be equal to $C_1$. Hence

$$\frac{1}{C_1} = \frac{1}{C_x} + \frac{1}{C_2}$$

$$C_x = \frac{C_1 C_2}{C_2 - C_1} \qquad [52]$$

Hence $C_x$ can be determined. The total resistance in the circuit is $R$ plus the leakage resistance $R_x$ which can be assumed to be in parallel with the unknown capacitor. Hence

$$R_x = \text{total resistance} - R$$

But for the short-circuit situation we have $R = 1/\omega_0 C_1 Q_1$ and for the situation where the unknown capacitor is in circuit we must have the total resistance $= 1/\omega_0 C_2 Q_2$. Hence

$$R_x = \frac{C_1 Q_1 - C_2 Q_2}{\omega_0 C_1 C_2 Q_1 Q_2} \qquad [53]$$

The Q-factor for the unknown capacitor $Q_x$ is $1/\omega_0 R_x C_x$ and so

$$Q_x = \frac{Q_1 Q_2 (C_1 - C_2)}{C_1 Q_1 - C_2 Q_2} \qquad [54]$$

Thus equations [52], [53] and [54] can be used to determine the capacitance, leakage resistance and Q-factor of the unknown capacitor.

If the unknown component had been a small inductor instead of large capacitor then a similar procedure can be adopted of tuning the circuit with the unknown inductor short-circuited and then in circuit. Then

$$L_x = \frac{C_1 - C_2}{\omega_0^2 C_1 C_2} \qquad [55]$$

If the unknown component had been a pure resistor then when a similar procedure is adopted and the circuit tuned and the Q-factor obtained with the unknown resistor short-circuited and then in circuit, $C_1$ would equal $C_2$ since there would have been no change in resonance condition. Hence equation [53] becomes

$$R_x = \frac{Q_1 - Q_2}{\omega_0 C_1 Q_1 Q_2} \qquad [56]$$

For high impedance components such as small capacitors (below about 400 pF), large inductors (above about 100 mH) and large resistors, the unknown component is connected in parallel with the variable capacitor (Fig. 9.31). With the

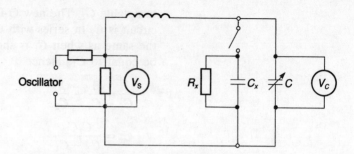

**Fig. 9.31** Measurement of a high impedance

unknown component open circuit, the circuit is tuned. At the resonance condition the variable capacitor is $C_1$ and the Q-factor $Q_1$. Then with the component connected into circuit the variable capacitor is adjusted to give resonance at the same frequency. If the variable capacitor is then $C_2$ and the Q-factor $Q_2$, and if the unknown has a capacitance of $C_x$ the parallel arrangement must give the same capacitance as when the unknown component is out of circuit and so

$$C_x + C_2 = C_1$$

Hence

$$C_x = C_1 - C_2 \qquad [57]$$

The leakage resistance $R_x$ of the unknown capacitor is in parallel with the resistance $R$ of the circuit, thus

$$\text{total resistance} = \frac{R_x R}{R_x + R}$$

But initially when the unknown capacitor is out of circuit we have $R = Q_1/\omega_0 C_1$ and when it is in circuit total resistance = $Q_2/\omega_0 C_2$. Hence

$$\frac{Q_2}{\omega_0 C_2} = \frac{R_x Q_1/\omega_0 C_1}{R_x + Q_1/\omega_0 C_1}$$

$$\frac{Q_2}{\omega_0 C_2} = \frac{R_x Q_1}{\omega_0 C_1 R_x + Q_1}$$

$$R_x = \frac{Q_1 Q_2}{\omega_0 C_1 (Q_1 - Q_2)} \qquad [58]$$

The Q-factor of the unknown capacitor $Q_x$ is

$$Q_x = \frac{Q_1 Q_2 (C_1 - C_2)}{C_1 (Q_1 - Q_2)} \qquad [59]$$

If the unknown is inductive $L_x$,

$$L_x = \frac{1}{\omega_0^2 (C_1 - C_2)} \qquad [60]$$

and its parallel resistance is given by equation [58].

In the above discussions for measurements with an inductor, whether it be in series or parallel in the circuit, it has been assumed that the inductor can be represented by an inductance either in series or in parallel with a resistance. Often a more realistic equivalent circuit for the inductor is also to include capacitance. Thus, for example, we could represent an inductor by the equivalent circuit shown in Fig. 9.32. This capacitance is often referred to as *self-capacitance*. This self-capacitance can be found by taking two measurements at different frequencies.

Using the arrangement of Fig. 9.29, the tuning capacitor is set to a high value, often its maximum value, and the resonance condition obtained by adjusting the oscillator frequency. If this frequency is $f_1$ when the tuning capacitor is $C_1$, then since the condition for series resonance is

$$f_1 = \frac{1}{2\pi\sqrt{LC}}$$

where $C$ is the total circuit capacitance due to $C_1$ and $C_x$. Since these can be considered to be effectively in parallel then

$$f_1 = \frac{1}{2\pi\sqrt{L(C_x + C_1)}}$$

The frequency is then set to $2f_1$ and resonance obtained by adjusting the tuning capacitor. If this has a value $C_2$ then

$$2f_1 = \frac{1}{2\pi\sqrt{L(C_x + C_2)}}$$

Thus

$$\frac{2}{2\pi\sqrt{L(C_x + C_1)}} = \frac{1}{2\pi\sqrt{L(C_x + C_2)}}$$

$$4(C_x + C_2) = C_x + C_1$$

$$C_x = \frac{C_1 - 4C_2}{3} \qquad [61]$$

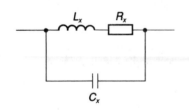

**Fig. 9.32** An inductor

### Example 12

A capacitor is being measured using a Q-meter in the parallel arrangement of Fig. 9.31. With the unknown capacitor out of circuit the tuning capacitor is set to 500 pF at a frequency of 220 kHz for resonance, the indicated Q-factor being 180. With the unknown capacitor in circuit the tuning capacitor has to be set to 320 pF for resonance at 220 kHz, the Q-factor being 176. What is the capacitance and leakage resistance of the capacitor?

*Answer*

Using equation [57]

$$C_x = C_1 - C_2 = 500 - 320 = 180\,\text{pF}$$

Using equation [58]

$$R_x = \frac{Q_1 Q_2}{\omega_0 C_1 (Q_1 - Q_2)} = \frac{180 \times 176}{2\pi \times 220 \times 10^3 \times 500 \times 10^{-12} \times 4}$$

$$= 11.5\,\Omega$$

**Example 13**

An inductor is being measured using the series arrangement of a Q-meter. At a frequency of 2 MHz resonance was obtained with the tuning capacitor set at 500 pF. At 4 MHz resonance was obtained with the tuning capacitor at 120 pF. What was the self-capacitance of the inductor?

*Answer*

Using equation [61]

$$C_x = \frac{C_1 - 4C_2}{3} = \frac{500 - 4 \times 120}{3} = 7.3\,\text{pF}$$

**Potentiometer measurement system** The potentiometer measurement system involves using a potentiometer to produce a variable potential difference which can then be used to balance, and so cancel out, the potential difference being measured. Figure 9.33 shows the basis of the system. The working battery is used to produce a potential difference across the full length of the potentiometer track. The potential difference across the length $L$ of the potentiometer track is then balanced against the unknown e.m.f., the length $L$ being adjusted until no current is detected by the galvanometer. When this occurs the unknown e.m.f. $E$ must be equal to the potential difference across the length $L$ of the track. If the track is uniform then

$$E = kL$$

where $k$ is a constant, in fact the potential difference per unit length of track. This can be determined by repeating the balancing operation with a standard cell of e.m.f. $E_s$. Then

$$E_s = kL_s$$

where $L_s$ is the balance length with the standard cell. Hence

$$E = \frac{E_s L}{L_s}$$

With a commercial form of the potentiometer measuring system the movement of the potentiometer slider over the

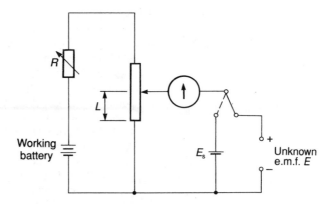

**Fig. 9.33** Potentiometric measuring system

track results in the movement of a pointer over a scale. This scale is calibrated directly in volts. The standardization is achieved by setting the pointer to the required value of the standard cell e.m.f. and then adjusting $R$ until balance occurs. The potentiometer measurement system does not depend on the calibration of a galvanometer since it is only used to indicate when there is zero current. The system is essentially an infinite impedance voltmeter.

The system can be used for the comparison of the resistances of two resistors. The two are connected in series so that the same current passes through them, then the potential differences across the two are compared by connecting the potentiometer across first one of the resistors and then across the other (Fig. 9.34). The ratio of the resistances is the ratio of the balance lengths.

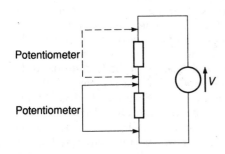

**Fig. 9.34** Comparing resistances

### Automatic bridges

The d.c., a.c., transformer bridges and potentiometers can be made automatic, i.e., self-balancing. This means that, in the case of a bridge, the out-of-balance current or potential difference is used, after amplification, to operate a motor that can move a slider across some variable resistance or other element until the out-of-balance current becomes zero. The position of the slider then gives a measure of the unknown quantity which was responsible for the out-of-balance current. An example of a self-balancing potentiometer is shown in Fig. 8.10.

### Problems

1 Neglecting loading, estimate the accuracy with which a resistance can be measured using the ammeter–voltmeter method if the meters used have accuracies of $\pm1\%$.

2 Explain why, when using the ammeter–voltmeter method for the measurement of resistance, a high resistance voltmeter is desirable if reasonable accuracy is to be obtained.

3 What factors limit the accuracy with which a resistance can be measured by the ammeter–voltmeter method?

4 A d.c. Wheatstone bridge has resistances of $20\,\Omega$ in arm BC, $500\,\Omega$ in arm CD and $200\,\Omega$ in arm AD. What will be the resistance in arm AB if the bridge is balanced?

5 A d.c. Wheatstone bridge has a 6.0 V supply connected between points A and C. What will be the potential difference between points B and D when the resistances in the bridge arms are AB $10\,\Omega$, BC $20\,\Omega$, CD $60\,\Omega$ and AD $31\,\Omega$?

6 A d.c. Wheatstone bridge has a 5.0 V supply connected between points A and C and a galvanometer of resistance $50\,\Omega$ between points B and D. What will be the current through the galvanometer when the resistances in the bridge arms are AB $120\,\Omega$, BC $120\,\Omega$, CD $120\,\Omega$ and DA $120.1\,\Omega$?

7 A Wheatstone bridge has a supply voltage of 5 V and negligible internal resistance and a galvanometer of current sensitivity $5\,\text{mm}/\mu\text{A}$ and resistance $100\,\Omega$. The bridge is initially balanced when the resistances in the arms are AB $1000\,\Omega$, BC $2000\,\Omega$, CD $200\,\Omega$, DA $100\,\Omega$. What will be the deflection given by the galvanometer when the resistance in arm BC changes by $5\,\Omega$?

8 Explain how the Wheatstone bridge can be used to determine the resistance of the coil of a platinum resistance thermometer and compensate for any resistance changes in the leads to the coil.

9 A platinum resistance thermometer has a resistance at 0 °C of $120\,\Omega$ and forms one arm of a Wheatstone bridge. At this temperature the bridge is balanced with each of the other arms also being $120\,\Omega$. The temperature coefficient of resistance of the platinum is $0.0039\,°\text{C}^{-1}$. What will be the output voltage for change in temperature of 20 °C if the instrument used to measure it can be assumed to have infinite resistance and the supply voltage, with negligible internal resistance, for the bridge is 6.0 V?

10 A Kelvin double bridge, of the form shown in Fig. 9.11, is used with $R_3/R_4 = r_3/r_4$. At balance, $R_4 = 100\,\Omega$, $R_2 = 10.5\,\Omega$ and $R_3 = 0.2\,\Omega$. What is the value of the resistance $R_1$?

11 For the basic a.c. bridge shown in Fig. 9.15 the impedances are $Z_2 = 200 \angle 20°\,\Omega$, $Z_3 = 100\,\Omega$, and $Z_4 = 50 \angle -60°\,\Omega$. What is the value of $Z_1$ at balance?

12 A Maxwell bridge is used to determine the inductance and resistance of an inductor. At balance the bridge arms are AB $1.0\,\mu\text{F}$ in parallel with $10\,\text{k}\Omega$, BC $400\,\Omega$, CD the inductor, and DA $500\,\Omega$. What are the inductance and resistance of the inductor?

13 A Hay bridge is used with a 1 kHz a.c. source and at balance the bridge arms are AB $0.1\,\mu\text{F}$ in series with $95\,\Omega$, BC $985\,\Omega$, CD the inductor and DA $500\,\Omega$. What are the value of the inductance and the series resistance?

14 An a.c. bridge has in arm AB a $1.2\,\text{k}\Omega$ resistor, in arm BC a $2.0\,\text{k}\Omega$ resistor in series with a $1\,\mu\text{F}$ capacitor, in arm AD an unknown capacitor having capacitance and series resistance and in arm DA a $0.5\,\text{k}\Omega$ resistance. What are the values of the unknown capacitance and series resistance?

15 A series capacitance bridge with an a.c. source of 1 kHz has at balance arms of AB 2 kΩ, BC 0.1 μF with 110 Ω, CD the unknown capacitor and DA 1 kΩ. What are the capacitance, series resistance and dissipation factor of the unknown capacitor?

16 A parallel capacitance bridge with an a.c. source of 1 kHz has at balance arms of AB 500 Ω, BC 0.10 μF in parallel with 430 Ω, CD the unknown capacitor and DA 1 kΩ. What are the capacitance, parallel resistance and dissipation factor of the unknown capacitor?

17 A single ratio transformer bridge with $N_1 = N_2$ is used to measure the parallel components of an unknown impedance at a frequency of 5 kHz. The bridge was balanced using standard capacitors and resistors and the scaled results obtained were $-105$ μF and 20 Ω. What are the parallel components?

18 Explain the principles of the double ratio transformer bridge and how it can be used to determine impedances.

19 Explain how a Q-meter can be used to measure the capacitance and leakage resistance of a capacitor of the order of 200 pF.

20 An inductor is being measured using the series arrangement of a Q-meter. At a frequency of 1 MHz resonance was obtained with the tuning capacitor set at 450 pF. At 2 MHz resonance was obtained with the tuning capacitor at 107 pF. What was the self capacitance of the inductor?

21 A capacitor is being measured using a Q-meter in the parallel arrangement of Fig. 9.31. With the unknown capacitor out of circuit the tuning capacitor is set to 500 pF at a frequency of 210 kHz for resonance, the indicated Q-factor being 140. With the unknown capacitor in circuit the tuning capacitor has to be set to 340 pF for resonance at 210 kHz, the Q-factor being 132. What is the capacitance and leakage resistance of the capacitor?

# 10 Power and energy measurement

**Introduction**

This chapter is about the measurement of power and energy in electrical circuits. In electrical circuits, whether d.c. or a.c., the instantaneous power delivered to a load is the product of the instantaneous current through the load and the potential difference across it. With a.c. circuits, because the current and potential difference are varying with time, a more useful quantity is the average power. This is the product of the root-mean-square values of the current $I$ and potential difference $V$ and the cosine of the phase angle $\phi$ between the current and potential difference, i.e.,

$$P = IV\cos\phi \qquad [1]$$

This power is often referred to as the *true power*, with $\cos\phi$ being called the *power factor*. When the current lags the voltage the power factor is said to be *lagging* and when the current leads the voltage *leading*. The product of the root-mean-square values of the current and potential difference is known as the *apparent power* or the volt–ampere product. Thus equation [1] can be written as

$$\text{true power} = \text{apparent power} \times \cos\phi \qquad [2]$$

This relationship can be represented by the *power triangle* (Fig. 10.1), with the apparent power being the hypotenuse of the triangle and the base the true power. The vertical side of the triangle is called the *reactive power Q* and is

$$\text{reactive power } Q = \text{apparent power} \times \sin\phi \qquad [3]$$

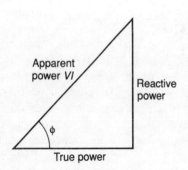

Apparent power *VI*

Reactive power

$\phi$

True power

**Fig. 10.1** Power triangle

The units of true power are watts, of apparent power volt-ampere (VA) and of reactive power Vars.

Instruments designed to measure the true power are called *wattmeters*, the most commonly used form being the dynamometer (see Chapter 6). The product of the true power and time is the total energy dissipated in a load in that time. Instruments designed to measure energy are called *watt-hour meters*.

## Single-phase wattmeter

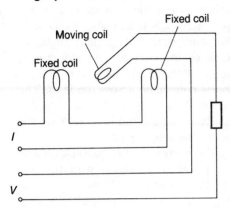

**Fig. 10.2** Wattmeter for single-phase load

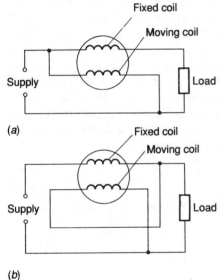

**Fig. 10.3** Wattmeter connections

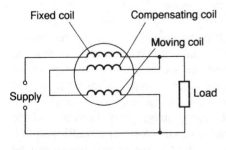

**Fig. 10.4** Compensated wattmeter

The principles of the dynamometer are discussed in Chapter 6. Basically the instrument consists of a moving coil in the magnetic field produced by a pair of fixed coils. When used as an ammeter or voltmeter the fixed and moving coils are connected in series. When used for the measurement of power dissipation (Fig. 10.2) the stationary coils are connected in series with the measurement circuit load and so the magnetic field produced is proportional to the current. The moving coil, with a series resistor, is connected in parallel with the load and so the current through it is a measure of the potential difference across the load. Since the angular deflection of the movable coil is proportional to the product of the currents in the fixed and movable coils, the angular deflection is proportional to the product of the current and potential difference. Theoretically the instrument responds to the instantaneous power but system inertia means that the instrument responds to the average power.

The wattmeter can be connected into a circuit in two ways (Fig. 10.3). In Fig. 10.3(a) the moving coil gives a measure of the potential difference across the fixed coils and the load, the fixed coils measuring the current through just the load. In Fig. 10.3(b) the potential difference is just that across the load and the current is that through the movable coil and the load. If $I_L$ is the load current, $R_m$ the resistance of the movable coil and $R_f$ the resistance of the fixed coils, then with circuit (a) the wattmeter reads high by the power due to the potential drop across the fixed coils, i.e., $I_L^2 R_f$ and with (b) the wattmeter reads high by the power due to the current through the movable coil, i.e., $V_L^2/R_m$. In both cases the wattmeter reads high. The error is minimized if circuit (a) is used for low current–high voltage loads and (b) for high current–low voltage loads.

This error can be overcome by using a compensated wattmeter (Fig. 10.4). With such a meter the fixed coils each have two windings with the same number of turns. One winding uses heavy gauge wire and carries the load current, the other with finer gauge wire is connected in series with the movable coil and thus carries the same current as that coil. This voltage coil current is, however, in the opposite direction to the load current through the fixed coils and cancels out the proportion of the magnetic flux due to the voltage coil current. The wattmeter thus indicates the correct power.

Wattmeters can be used for d.c. or a.c. power measurements up to about 2 or 3 kHz, being unaffected by the waveform of the voltage or current. They can be used with voltages up to about 300 V and currents of 20 A, with an accuracy of about ±0.1 to 0.5% of full-scale deflection.

## Power factor measurement

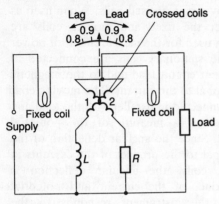

**Fig. 10.5** Crossed coil power factor meter

A form of dynamometer can be used for the measurement of the power factor, being then called the *crossed coil power factor meter*. Such a dynamometer has the moving element as two coils, mounted on the same shaft but at right angles to each other (Fig. 10.5). One of the coils has an inductor in series with it and is connected across the load, the other a resistor in series and it also is connected across the load. The currents in the two coils are equal in magnitude but displaced in time by 90°. The fixed coils are connected in series with the supply line and so respond to the line current.

For a power factor of 1 the current in the movable coil in series with the inductor (1) will be in phase with the line current and the current in the other coil (2) will be out of phase by 90°. Consequently coil 1 will rotate until it is in a plane parallel to that of the fixed coils. For a power factor of 0 the current in the movable coil in series with the resistor (2) will be in phase with the line current and the current in the other coil (1) will be out of phase by 90°. Consequently coil 2 will now rotate until it is in a plane parallel to that of the fixed coils. For other power factors the deflection of the coils will be to some intermediate position and thus a measure of the power factor. The deflection of the shaft on which the coils are mounted results in the movement of a pointer across a scale. The scale gives direct readings of the power factor, there being a central reading of 1 with lagging values decreasing to one side and leading values to the other.

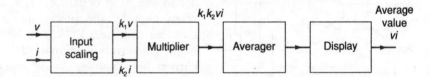

**Fig. 10.6** Electronic voltmeter

## Electronic wattmeter

With the electronic wattmeter the processes corresponding to the multiplication of the load current and potential difference and the taking of an average are carried out electronically, Fig. 10.6 summarizing the process. This process may be continuous or employ sampling. There are a number of methods used to give multiplication with a continuous display of the power. With one method the current determines the width of a pulse and the voltage the height. The average pulse area over a period of time thus becomes a measure of the average power. With the sampling method samples are simultaneously taken of the voltage and current. These samples are then converted to digital form and multiplied and averaged by digital circuits. The electronic wattmeter can be used over a range of about 0.1 W to 100 kW at frequencies up to 100 kHz. Its accuracy tends to decrease with frequency, from about ±0.5% at low frequencies to ±1% at high frequencies.

**Three-phase wattmeter**

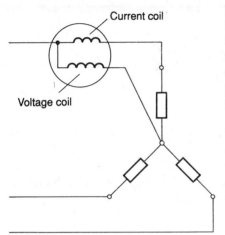

**Fig. 10.7** Measurement of power with a balanced load

When there is a balanced load with a three-phase system, it is possible to measure the total power consumed by using just a single wattmeter. With a balanced load the total power is the sum of the power consumed per load element. Figure 10.7 shows how a wattmeter can be used to make such a measurement with a star-connected balanced load. The total power consumed by the load is thus three times the instrument reading.

The power in any three-phase system, whether with a balanced or unbalanced load, can be measured using two or three wattmeters. For a star-connected balanced, or unbalanced, load a wattmeter may be connected to each load element and the total power thus becomes the sum of the three wattmeter readings (Fig. 10.8).

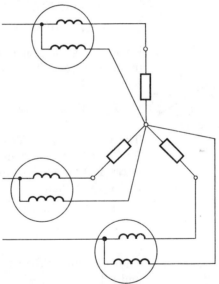

**Fig. 10.8** Three wattmeter method

Figure 10.9 shows how two wattmeters may be used to determine the power in a three-phase balanced, or unbalanced, three-wire circuit. For that circuit when balanced, the power reading indicated by wattmeter 1 is

$$P_1 = V_{RB}I_R \cos(\text{angle between phasors } \mathbf{I_R} \text{ and } \mathbf{V_{RB}})$$

The voltage $\mathbf{V_{RB}}$ is $\mathbf{V_{RS}} - \mathbf{V_{BS}}$ and is indicated by the phasor diagram in Fig. 10.10. For a balanced system the magnitudes of $\mathbf{V_{RS}}$ and $\mathbf{V_{BS}}$ are the same and so the phase angle between $\mathbf{V_{RB}}$ and $\mathbf{V_{RS}}$ is 30°. If the load has a lagging phase angle $\phi$,

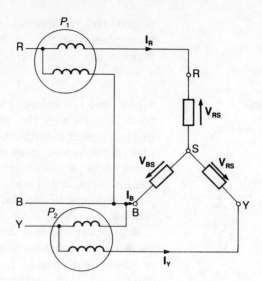

**Fig. 10.9**  Two wattmeter method

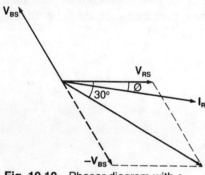

**Fig. 10.10**  Phasor diagram with a lagging power factor

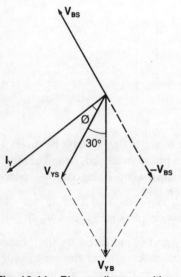

**Fig. 10.11**  Phasor diagram with a lagging power factor

i.e., the phase angle between $I_R$ and $V_{RS}$ is $\phi$, then the phase angle between $I_R$ and $V_{RB}$ is $(30° - \phi)$. Hence

$$P_1 = V_L I_L \cos(30° - \phi)$$

$$= V_L I_L (0.87 \cos\phi + 0.5 \sin\phi) \qquad [4]$$

where $V_L$ is the magnitude of the line voltage and $I_L$ that of the line current. The power reading indicated by wattmeter 2 is

$$P_2 = V_{YB} I_Y \cos(\text{angle between } I_Y \text{ and } V_{YB})$$

The voltage $V_{YB}$ is $V_{YS} - V_{BS}$ and is indicated by Fig. 10.11. For a balanced system we thus have a phase angle of 30° between $V_{YB}$ and $V_{YS}$ and since the phase angle between $V_{YS}$ and $I_Y$ is $\phi$ then

$$P_2 = V_L I_L \cos(30° + \phi)$$

$$= V_L I_L (0.87 \cos\phi - 0.5 \sin\phi) \qquad [5]$$

Thus the sum of the wattmeter readings is, using equations [4] and [5],

$$P_1 + P_2 = V_L I_L (0.87 \cos\phi + 0.5 \sin\phi$$
$$+ 0.87 \cos\phi - 0.5 \sin\phi)$$

$$= \sqrt{3}\, V_L I_L \cos\phi \qquad [6]$$

This is the total power that would be consumed by a balanced load. Thus

$$\text{total power} = P_1 + P_2 \qquad [7]$$

i.e., the sum of the two meter readings is the total power consumed.

The difference between the two wattmeter readings is, using equations [4] and [15],

$$P_1 - P_2 = V_L I_L (0.87 \cos \phi + 0.5 \sin \phi$$
$$- 0.87 \cos \phi + 0.5 \sin \phi)$$

$$= V_L I_L \sin \phi \qquad \qquad [8]$$

For a three-phase system, the reactive power

$$Q = \sqrt{3} I_L V_L \sin \phi$$

Thus

$$Q = \sqrt{3} (P_1 - P_2) \qquad \qquad [9]$$

The value of the power factor $\cos \phi$ can be obtained as follows. Dividing equation [8] by equation [6] gives

$$\frac{\tan \phi}{\sqrt{3}} = \frac{P_1 - P_2}{P_1 + P_2} \qquad \qquad [10]$$

But $\cos^2 \phi + \sin^2 \phi = 1$, hence when this is divided by $\cos^2 \phi$

$$1 + \tan^2 \phi = \frac{1}{\cos^2 \phi}$$

Hence

$$1 + 3 \left( \frac{P_1 - P_2}{P_1 + P_2} \right)^2 = \frac{1}{\cos^2 \phi}$$

$$\cos \phi = \frac{1}{\sqrt{\left\{ 1 + 3 \left[ \frac{(P_1 - P_2)}{(P_1 + P_2)} \right]^2 \right\}}} \qquad \qquad [11]$$

The power factor $\cos \phi$ can thus be obtained from the two wattmeter readings.

### Example 1

Two wattmeters are used to measure the power consumed by a three-wire, balanced load system (as in Fig. 10.9). The meters gave readings of 50 kW and −30 kW. What is the total power consumed and the power factor?

*Answer*

The total power consumed is the sum of the meter readings (equation [7]). Thus the total power is $50 - 30 = 20$ kW. The power factor is given by equation [11] as

$$\cos \phi = \frac{1}{\sqrt{\left\{ 1 + 3 \left[ \frac{(P_1 - P_2)}{(P_1 + P_2)} \right]^2 \right\}}}$$

$$= \frac{1}{\sqrt{\left\{ 1 + 3 \left[ \frac{(50 + 30)}{(50 - 30)} \right]^2 \right\}}} = 0.14$$

## Watt-hour meter

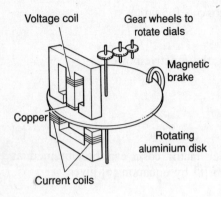

Voltage coil

Gear wheels to rotate dials

Magnetic brake

Copper

Rotating aluminium disk

Current coils

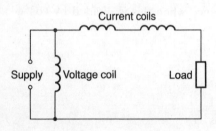

Current coils

Supply Voltage coil Load

**Fig. 10.12** Watt-hour meter

The watt-hour meter is widely used for the measurement of the electrical energy supplied to industrial and domestic consumers through the mains alternating supply. The most commonly used form is the shaded pole induction meter (Fig. 10.12). Alternating currents in the voltage and current coils of the meter produce alternating magnetic fields which interact with an aluminium disk and induce eddy currents in it. The magnetic flux generated by the voltage coil is 90° out of phase with that produced by the current coil. This is because the voltage coil magnetic circuit is capped by a copper sheet, hence the term shaded pole. The result of this is that the aluminium disk experiences a torque which causes it to rotate. The average torque acting on the disk is proportional to $VI\cos\phi$:

$$\text{generated torque} = k_g VI \cos\phi$$

where $k_g$ is a constant, $V$ the voltage, $I$ the current and $\cos\phi$ the power factor. This torque is opposed by a magnetic brake. This is just a permanent magnet which interacts with the moving aluminium disk to produce a torque opposing the disk motion. This torque is proportional to the speed of rotation of the disk:

$$\text{braking torque} = k_b N$$

where $k_b$ is a constant and $N$ the number of revolutions per unit time. At equilibrium the generated and braking torques are equal and thus

$$k_b = k_g VI \cos\phi$$

The speed of rotation of the disk is thus proportional to $VI\cos\phi$, i.e., the true power. The shaft of the disk is connected via gearing to a mechanical counter which thus gives a count proportional to the watt-hours.

## Problems

1  Explain how the electrodynamometer can be used as a wattmeter.
2  Explain the principle of the compensated wattmeter and why it is necessary for such compensation.
3  An electrodynamometer wattmeter gives a full-scale reading of 100 W. What will be the half-scale reading?
4  Two wattmeters are used to measure the power consumed by a three-wire, balanced load system (as in Fig. 10.9). The meters gave readings of 7 kW and 3 kW. Find the total power consumed and the power factor.
5  Describe the principles of operation of the domestic electrical meter which is used to determine the amount of electrical energy used in a household.
6  An electronic wattmeter includes in its specification the term conversion rate 600 ms. Explain the significance of the term.

# 11  Oscilloscopes

The cathode ray oscilloscope is one of the most commonly used instruments in electronics. It works on the principle of using a beam of electrons to 'paint' a display on the phosphor-coated screen of a cathode ray tube. The display is in the form of a two-dimensional graph showing how the voltage of a signal varies either with time or with some other signal. The oscilloscope is essentially just a form of voltmeter. Additional components, however, enable the oscilloscope to be used for more than just a voltage display.

This chapter is a consideration of the basic form of oscilloscopes and their subsystems, and an indication of the types of measurements that they can be used for. Oscilloscopes can be considered to fall into two main categories: real time and non-real time. *Real time* oscilloscopes give screen displays of what is actually happening in some circuit and are virtually restricted to the observation of rapidly repetitive waveforms. *Non-real time* oscilloscopes allow the observation of non-repetitive waveforms since data about the waveform are either stored or sampled and the waveform then reconstructed at some later time.

**The basic oscilloscope**

The basic oscilloscope can be represented as being composed of a number of interconnected subsystems, as in Fig. 11.1. These subsystems are:

1   The *display system*, i.e., the cathode ray tube, which converts vertical and horizontal deflection signals into a deflection of a fluorescent spot on a screen.
2   The *vertical deflection system* which results in an input producing a deflection on the display screen in the vertical direction.
3   The *horizontal deflection system* which results in an external input producing a deflection on the display screen in the horizontal direction or an internal timebase signal a

deflection at a constant speed across the screen in the horizontal direction.

4  *Power supplies*.

**The cathode ray tube**

Figure 11.2 shows the basic features of the *cathode ray tube*. Electrons are produced by the heating of the cathode. The number of these electrons which form the electron beam, i.e., its brilliance, is determined by a potential applied to an electrode, the modulator, immediately in front of the cathode. Some oscilloscopes have an input labelled $Z$ modulation which allows the brightness of the beam to be controlled by an external signal. The electrons are accelerated down the tube by the potential difference between the cathode and the anode. An electron lens is used to focus the beam so that when it reaches the phosphor-coated screen it forms a small luminous spot. The focus is adjusted by changing the potential of the electrodes relative to that of the earlier electrodes. The electron emission, modulator, anode and lens is known as the electron gun. The beam can be deflected in the vertical ($Y$)

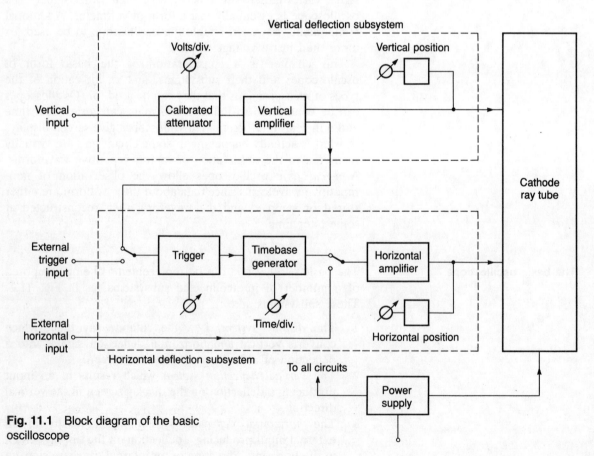

**Fig. 11.1**  Block diagram of the basic oscilloscope

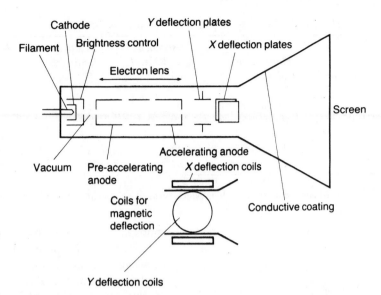

**Fig. 11.2** Cathode ray tube

direction by a potential difference applied between the $Y$ deflection plates. A potential difference between the $X$ deflection plates will cause it to deflect in the horizontal ($X$) direction. In some cathode ray tubes the beam is deflected by the $Y$ and $X$ inputs being applied to coils on either side of the tube and so deflecting the beam as a result of their magnetic fields. Electrostatic deflection systems are limited to giving deflections of about 40°, magnetic deflection systems can, however, give deflections up to about 110°. After deflection there is sometimes a further acceleration system. Such post-deflection acceleration is used because with an electrostatic deflection system high sensitivity requires the electron velocity to be low in passing between the plates. However, if a bright spot on the screen is to be obtained it is necessary for the electrons to have high velocities.

The *phosphor* used as the screen coating material emits light when hit by electrons. The light takes a little time to build up when the electron beam first hits the phosphor, this light emitted during the excitation of the phosphor being called fluorescence. When the electron beam stops bombarding the phosphor the light continues to be emitted for some time before decaying to zero, this light being called phosphorescence. The time taken for the light output to fall to some specified value of its initial value is known as the *decay time* or *persistence*. The time taken from the beginning of the excitation of the phosphor to it reaching 90% of the maximum intensity plus the persistence time is called the *writing time*. For a screen to give a flicker-free display it is necessary for the phosphor to be refreshed by further bombardment of electrons before the end of its decay time. Thus a short persistence

**Table 11.1** Characteristics of commonly used phosphors

| Phosphor | Fluorescence colour | Persistence to 10% level (ms) | Relative luminance (%) | Relative writing speed (%) | Comment |
|---|---|---|---|---|---|
| P1 | yellow-green | 95 | 50 | 20 | Mostly replaced by P31 |
| P4 | white | 120 | 50 | 40 | TV displays |
| P7 | blue* | 1500 | 35 | 75 | Long delay screens |
| P11 | blue | 20 | 15 | 100 | Photographic |
| P31 | yellow-green | 32 | 100 | 50 | General purpose |

\* P7 has a fluorescent colour of blue which turns to yellow-green phosphorescence.

phosphor requires more frequent refreshes than a long persistence phosphor. Long persistence phosphors will, however, result in trace afterglows. A phosphor used widely with oscilloscopes is P31. It gives a yellowish green trace and a decay time to 0.1% of 32 ms. Where the oscilloscope is used with a camera for photographing high speed traces P11 is used. It gives a blue trace with a decay time to 0.1% of 20 ms. Table 11.1 shows typical characteristics of some commonly used phosphors.

**Vertical deflection subsystem**

A signal applied to the *Y deflection plates* causes the electron beam, and hence the spot on the screen, to move up or down in the vertical direction. The vertical deflection subsystem consists of an input selector, a switched attenuator, a preamplifier, the main amplifier and a delay line (Fig. 11.3). The input selector passes the input signal to the attenuator and enables a d.c. input signal to be passed directly to the attenuator, with an a.c. input signal being passed through a capacitor which blocks off any d.c. components. The switched attenuator is used to change the magnitude of the signal fed to the amplifier and so obtain different deflection sensitivities. It needs to give a constant loading at all settings and attenuate all frequencies equally. The input impedance is typically about 1 MΩ shunted by a capacitance of about 10–100 pF. The amplifiers are used to provide a constant gain. These, together with the switched attenuator, result in sensitivities which, for a general purpose oscilloscope, vary between 5 mV per scale division to 20 mV per scale division. Accuracy with vertical measurements depends to a large extent on the linearity of the amplifiers and is usually about ±1–3%. When the system is in its a.c. mode its bandwidth typically extends from about 2 Hz to 10 MHz; when in its d.c mode from d.c. to 10 MHz. Because there is some delay in the start of the horizontal sweep of the electron beam, it is necessary to delay the vertical system input

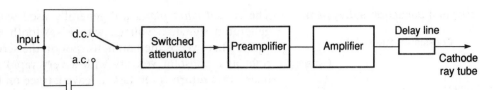

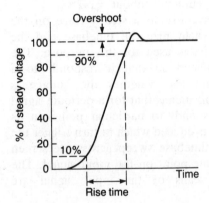

**Fig. 11.3**  Vertical deflection subsystem

**Fig. 11.4**  Rise time and overshoot

to the cathode ray tube. This is done by a delay line, this generally consisting of a coil wound on a core and behaving as a line consisting of series inductance and parallel capacitance elements.

Figure 11.4 shows the type of response that is obtained when a step input, i.e., an abrupt change in input voltage, is applied to the $Y$ deflection system. Some time elapses before the deflection indicates 100% of the applied voltage, with some overshoot occurring before the steady 100% value occurs. The term *rise time* is used for the time taken for the deflection to go from 10% to 90% of its steady deflection. In general, the product of the rise time and bandwidth is a constant in the range 0.25–0.5. For minimum overshoot the optimum value is 0.35. Thus a bandwidth of 10 MHz would mean that there was a rise time of about $35 \times 10^{-9}$ s (35 ns):

$$\text{Bandwidth} \times \text{rise time} \approx 0.35 \qquad [1]$$

The effect of the oscilloscope rise time on a measurement is to give a false reading of the real rise time of the input signal. Thus if the signal has a rise time $t_s$ and the oscilloscope a rise time $t_o$ then the indicated signal has a rise time $t_i$ given by

$$t_i^2 = t_s^2 + t_o^2 \qquad [2]$$

The error of the indicated rise time is the difference between the indicated rise time and the actual signal rise time. With the signal rise time about twice the oscilloscope rise time the error is about 10%. For errors less than 2% the signal rise time has to be five times that of the oscilloscope.

**Example 1**

The $Y$ deflection system of an oscilloscope has a bandwidth of d.c. to 15 MHz. What would be the expected rise time?

*Answer*

Using equation [1]

$$\text{bandwidth} \times \text{rise time} \approx 0.35$$

then the rise time is about

$$0.35/(15 \times 10^6) = 23 \times 10^{-9}\,\text{s}$$

i.e., 23 ns.

### Horizontal deflection subsystem

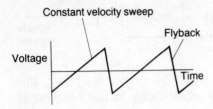

**Fig. 11.5** Sawtooth waveform

The *X deflection plates* are generally used with an internally generated signal, a voltage with a sawtooth waveform (Fig. 11.5). This sweeps the luminous spot on the screen from left to right at a constant velocity with a very rapid return, i.e., fly back. This return is too fast to leave a trace on the screen. The constant velocity movement from left to right means that the distance moved in the *X* direction is proportional to the time elapsed. Hence the sawtooth waveform gives a horizontal time axis, i.e., a *timebase*. Typically an oscilloscope will have a range of timebases from about 1 s per scale division to 0.2 μs per scale division with an accuracy of about ±1–2%.

For an input signal to give rise to a steady trace on the screen it is necessary to synchronize the timebase and the input signal. A *trigger circuit* is used for this purpose. The trigger circuit can be adjusted so that it responds to a particular voltage level and also whether the voltage is increasing or decreasing. This means that for a periodic signal input the trigger circuit responds to particular points in its cycle (Fig. 11.6). Pulses are produced which in turn trigger the timebase into action. The timebase sweep across the screen thus always starts at the same point on the input signal. The result is that successive scans of the input signal are superimposed.

The source of the signal used to operate the trigger circuit can be internal, using the output of the vertical amplifier and so making the input signal control the triggering. Alternatively the line signal, i.e., 50 or 60 Hz, or an external signal can be used. The line signal is useful if the vertical input signal is of the same frequency.

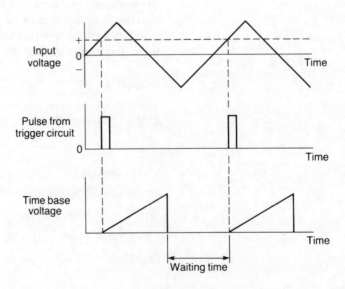

**Fig. 11.6** Triggering

**Oscilloscope probes**

While oscilloscope inputs can be connected to the circuit being measured by means of just a length of wire, this has the disadvantages that it may pick up interference and also that it can upset the normal function of the circuit because of the loading imposed by the oscilloscope. This loading is typically of the order of $1\,M\Omega$ shunted by $10–100\,pF$. While the use of coaxial cable to make the connections can protect against the pick-up of interference signals, it does not avoid the loading problem. Indeed the extra capacitance of the coaxial cable will increase the capacitive loading. A low capacitance cable of length $1\,m$ can add about $60\,pF$ and result in an input capacitance of the order of $100\,pF$ or more. Such a capacitance has a reactance as low as $160\,\Omega$ if the frequency is $10\,MHz$ and so the loading effects can be very significant. It is desirable, particularly for high frequency measurements, for the capacitive loading to be less than $20\,pF$.

To overcome these problems, *oscilloscope probes* are used. These are required:

1    to transmit an accurate representation of the circuit signal from the probe tip to the oscilloscope input;
2    to prevent the oscilloscope unduly loading the signal source and consequently giving loading errors;

and may also be required:

3    to attenuate or amplify the signal;
4    to convert the signal into a form which can be measured by the oscilloscope.

In general, probes consist of three elements: a probe head, the interconnecting cable and the cable termination (Fig. 11.7). The probe head contains the signal-sensing circuit and may be passive, only containing such circuit elements as resistors and capacitors, or active and containing such elements as amplifiers with field-effect transistors. A coaxial cable is used to transmit the signal from the probe head to the termination circuit. Termination circuits are not always included and thus the cable may connect directly with the cathode ray oscilloscope input. However the cable is a transmission line. Signals sent along a transmission line can reflect at the end and so produce signal distortion. Such reflections can be avoided if the cable is terminated with the characteristic impedance of the cable. The function of the termination circuit is to terminate the connecting cable in the characteristic impedance of the cable and so present the cable impedance to the oscilloscope input.

**Fig. 11.7**   Oscilloscope probe

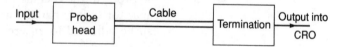

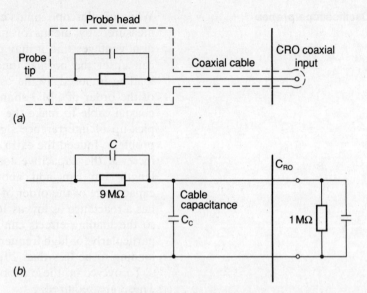

**Fig. 11.8** 10-to-1 probe

A variety of probes has been developed for general and specialized applications. The *passive voltage probe* is a very commonly used one. Such a probe attenuates, and attenuation ratios of 1-to-1, 10-to-1 and 100-to-1 are common. Figure 11.8(*a*) shows the form of a 10-to-1 unterminated probe and Fig. 11.8(*b*) the effective circuit when the cable capacitance and the input resistance and capacitance of the oscilloscope are taken into account. If the input signal is d.c. then the capacitors act as open circuits and so, to produce a potential difference across the oscilloscope input which is 1/10th of the input to the probe head, we need a resistance of nine times the oscilloscope input resistance. If the input resistance is $1\,\text{M}\Omega$ then the probe head has to include a resistance of $9\,\text{M}\Omega$. When the input to the probe head is a.c., the reactances of the capacitive elements must also be in this ratio 9-to-1. The cable capacitance $C_c$ is in parallel with that of the oscilloscope input capacitance $C_o$ and so the total capacitance across the input is $C_c + C_o$. This has a reactance of $1/\omega(C_c + C_o)$ and so we must have

$$\frac{9}{1} = \frac{1/\omega C}{1/\omega(C_c + C_o)}$$

$$\frac{9}{1} = \frac{C_c + C_o}{C} \qquad [3]$$

Because the capacitance of the oscilloscope can vary from one instrument to another, a variable capacitor is often included in the probe at the oscilloscope end of the cable (Fig. 11.9). This capacitor $C_t$ is in parallel with the oscilloscope capacitance and can be adjusted to ensure the 9-to-1 ratio of reactances. The

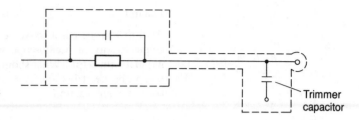

**Fig. 11.9** 10-to-1 probe with trimmer capacitor

Trimmer capacitor

**Fig. 11.10** Effect of (a) under, (b) correct and (c) overcompensation on a square waveform

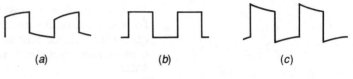

(a)    (b)    (c)

correct setting of the probe capacitors can be checked by observing a square wave input to the probe. When the capacitances are correctly set the square wave has square corners (Fig. 11.10(b)). If $(C_t + C_c + C_o)$ is less than $9C$ then the corners of the square waveform are rounded (Fig. 11.10(a)); if greater, the corners show overshoot (Fig. 11.10(c)).

*Active probes* tend to have a small amplifier using field-effect transistors built into the probe head. This enables the probe and oscilloscope to present a high impedance to the input signal source, typically about $10\,M\Omega$ shunted by $0.5\,pF$, and makes them very useful where long cable runs are necessary. However, the field-effect transistors limit the dynamic range of input signal that can be handled to between 0.5 and 5 V. The term dynamic range in this context means the maximum peak-to-peak signal that can be accepted. To extend this range, add on 10-to-1 and 100-to-1 attenuators can be used.

*Current probes* allow the measurement of alternating current in a conductor without the need to insert any component into the circuit for which the current is being measured. Such a probe is clamped round the conductor, as in Fig. 11.11. The current-sensing element is a transformer with a magnetic core around which a secondary coil is wound and the current-carrying conductor acts as the primary coil. The output from the secondary coil is amplified and typically gives about 1 mV per mA. It can only be used with alternating current and has a frequency range of about 100 Hz to 100 MHz. Another version of the current probe uses the Hall effect to sense the magnetic field produced in a magnetic core by the current in the conductor. Such a current probe can be used with d.c. and has a range from d.c. to about 50 MHz.

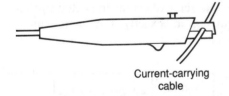

Current-carrying cable

**Fig. 11.11** Current probe

### Example 2

A 10-to-1 probe has a coaxial cable with a capacitance of 60 pF and is connected to an oscilloscope with an input impedance of 1 MΩ in parallel with 30 pF. What should be the resistance and capacitance in the probe head if there is no trimmer capacitor and there is to be correct compensation?

*Answer*

The total capacitance across the oscilloscope input is 60 + 30 = 90 pF. For correct compensation this must be nine times the probe head capacitance. Hence this must be 10 pF. The probe head resistance must be nine times the oscilloscope input resistance and so must be 9 MΩ.

**Two input oscilloscopes**

The terms dual trace and dual or double beam are used for oscilloscopes that enable two inputs to result in two traces on the screen. The term *channel* is used for each input and such oscilloscopes are said to be two channel. Two channels are the most common, though four and eight channel instruments are also available. *Double beam oscilloscopes* have two independent electron gun assemblies and so two electron beams. Each beam has its own Y deflection plates and can have either a common set of X deflection plates, and so a common timebase, or separate X deflection plates and so independent timebases (Fig. 11.12). A cheaper, more commonly used version is the *dual trace instrument* which uses a single electron gun and switches the Y deflection plates from one input signal to the other each time the timebase is triggered (Fig. 11.13). The display thus alternates between channels. This is known as the *alternate mode*. It has, however, the disadvantage that the two events indicated by the two traces are not occurring at exactly the same time. However, if the two events are cyclical this presents no

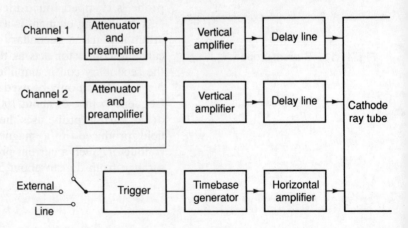

**Fig. 11.12** Dual beam oscilloscope

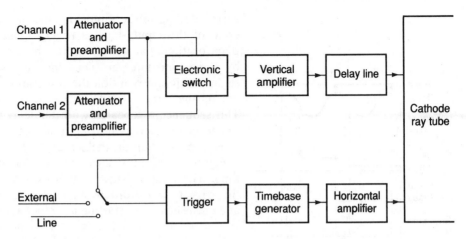

**Fig. 11.13** Dual trace oscilloscope

problems. Another alternative is to sample the two inputs more frequently, chopping from one channel to the other at a high frequency. This is known as the *chopped mode*. The frequency with which the deflection plates are switched from one signal to the other is typically about 150 kHz.

## Sampling oscilloscope

The *sampling oscilloscope* does not deal with an input signal in real time but takes samples of the input signals at different parts of its waveform on successive cycles and then assembles these to form a picture of the complete waveform. This technique enables it to cope with high speed signals and gives a very wide bandwidth, up to about 20 GHz. There is, however, the disadvantage that the waveform must be repetitive.

Figure 11.14 shows a block diagram of the sampling oscilloscope. The input signal is applied to the trigger, delayed, and then sampled by the sampling gate. The sample is then stored in the capacitative store and fed to the vertical plates of the cathode ray tube. It results in a dot on the screen

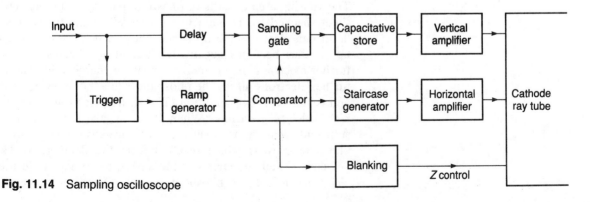

**Fig. 11.14** Sampling oscilloscope

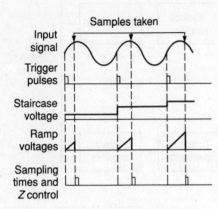

**Fig. 11.15** Sampling oscilloscope waveforms

corresponding to the sampled value of the input. On the next cycle of the input signal a trigger pulse occurs at the same point on the input waveform but the sample taking is delayed by some time interval. Thus the next sample is taken at a slightly different point on the waveform. The result is another dot on the screen corresponding to the second sample point. This sequence is repeated with the delay time from the trigger signal being increased each time and so a picture of the waveform built up on the screen.

Figure 11.15 illustrates how the waveforms change with time. The time at which samples are taken is determined by the trigger circuit producing pulses at regular time intervals and each such pulse starting a ramp voltage. This ramp voltage is fed to the comparator and compared with a steadily increasing staircase signal. When the comparator determines the two signals are equal it activates the sampling gate and the sample is taken. At the same time the comparator uses the $Z$ control of the cathode ray tube to brighten the screen for a short interval of time. The height of the staircase voltage at this time also determines the horizontal position on the screen.

**Storage oscilloscopes**

With *storage oscilloscopes* the trace produced by the $Y$ deflection plates remains on the screen after the input signal has ceased, only being removed by a deliberate action of erasure. There are two forms of analogue storage tube, the bistable storage and the variable persistence. The bistable tube is slower than the variable persistence tube; however, it is capable of much longer storage times. Variable persistence tubes have storage times which can be varied from a few milliseconds to several hours, bistable tubes can store waveforms for many hours.

Figure 11.16 shows the basic features of the *bistable storage tube*. The tube has three electron guns. Two of the guns, called the flood electron guns, are on all the time and permanently flood the viewing area with low velocity electrons. The viewing area consists of phosphor particles on a dielectric sheet, backed by a conducting layer. When the low velocity flood gun electrons fall on a phosphor particle they charge it up. It thus becomes negatively charged and begins to repel further electrons. It thus reaches a stable charge value and no further electrons hit it. The phosphor is in a 'not glowing' state and remains in that condition. The writing electron gun emits high velocity electrons. When it is on, the electrons have sufficient velocity to overcome the negative charge on a phosphor particle which resulted from the flood guns. The velocity is high enough for the electrons to knock further electrons out of the phosphor particle. These electrons are

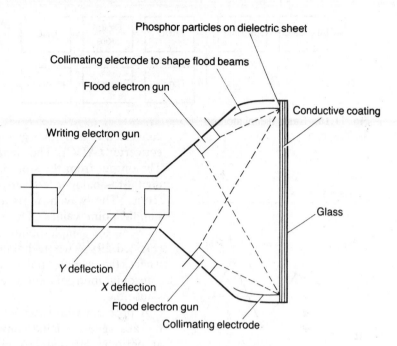

Phosphor particles on dielectric sheet

Collimating electrode to shape flood beams

Flood electron gun

Writing electron gun

Conductive coating

Glass

Y deflection

X deflection

Flood electron gun

Collimating electrode

**Fig. 11.16** Bistable storage tube

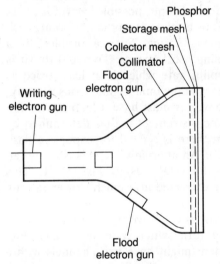

Phosphor

Storage mesh

Collector mesh

Collimator

Flood electron gun

Writing electron gun

Flood electron gun

**Fig. 11.17** Variable persistence storage tube

gathered by the conducting layer which backs the phosphor coated sheet. The result of the phosphor particles losing electrons is that they become positively charged. This charge remains, even when the writing gun stops emitting electrons. This is because the phosphor particle is being bombarded by flood electrons which are accelerated towards it by the positive charge. The acceleration is sufficient for secondary emission to continue. Thus the phosphor is in a 'glowing state' and remains in that condition.

Figure 11.17 shows the basic features of the *variable persistence tube*. The storage mesh consists of a thin layer of material such as magnesium fluoride deposited on a mesh. When electrons strike this mesh the areas where they strike become positively charged as a result of secondary emission. This charged state is maintained because of the action of the flood guns. Electrons from the flood guns also pass through the positively charged areas of the mesh and cause the phosphor behind to glow, so displaying the original waveform. To erase the picture the storage mesh is momentarily raised to the collector mesh potential.

## Digital storage oscilloscopes

Digital storage oscilloscopes use conventional cathode ray tubes but the input signals are digitized and storage occurs in electronic digital memories. Figure 11.18 shows the basic form of such an instrument. The analogue input signal, after attenuation and amplification, is sampled and the sample

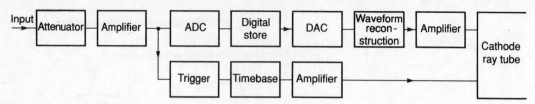

**Fig. 11.18** Basic digital storage oscilloscope

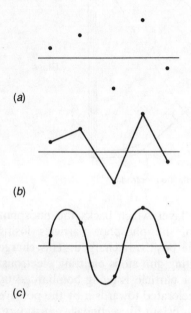

(a)

(b)

(c)

**Fig. 11.19** (a) With no interpolation, (b) linear interpolation, (c) sinusoidal interpolation

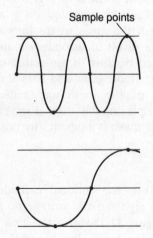

**Fig. 11.20** Aliasing

converted to a digital signal by means of an analogue-to-digital converter (ADC). The signal is then stored in a digital store. The output from the store is converted to analogue form by a digital-to-analogue converter (DAC) and fed to the oscilloscope. The waveform is reconstructed from a number of sample point values. The oscilloscope is able to interpolate between the sample points in order to give a continuous trace (Fig. 11.19). Two techniques are used, linear interpolation where the sample points are joined by straight lines and sinusoidal interpolation where the waveform is assumed to be sinusoidal.

A problem with the above basic form of instrument is that the analogue-to-digital converter samples the analogue signal at periodic intervals. A given set of sample points might, however, lead to more than one possible waveform, as indicated in Fig. 11.20. Thus a false display might occur, such a display being said to be *aliasing*. Such a problem is a characteristic of all sampling oscilloscopes. To avoid this it is necessary to have a sampling rate which is at least twice as high as the highest frequency in the input signal. Consequently, the analogue-to-digital converter must have a high conversion rate. The bandwidth of the instrument is thus determined by the ADC. This can mean there is a need for expensive flash converters (see Chapter 7). An alternative to this is to use an analogue store prior to the ADC, as in Fig. 11.21. This enables the samples to be extracted at a much slower rate by the ADC.

Digital storage oscilloscopes can have many input channels with the inputs being in turn switched to the ADC (Fig. 11.21). A typical instrument might have two channels with a bandwidth of 100 MHz, though instruments with as many as 40 channels are available. They can give an output which can be used to provide a hard copy of the stored data on an *X-Y* plotter and also an output which can be fed to a computer for processing. The oscilloscope can provide a range of display modes. In the *roll mode* the data and display are continually updated and so the screen presents a display, rather like a chart recorder would give, with a fixed pen and the paper being moved at a steady rate from right to left, the 'pen' in this case being just off screen to the right. The display thus scrolls across the screen as new data are introduced at the right-hand

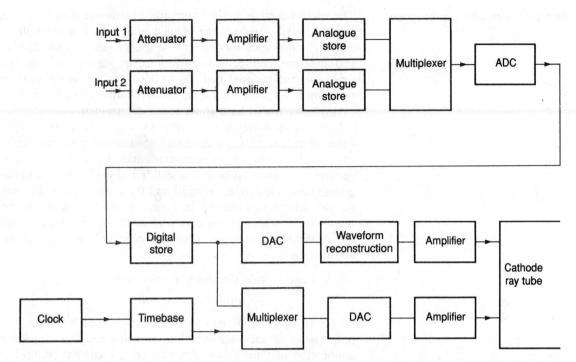

**Fig. 11.21** Multichannel digital
storage oscilloscope

edge. In the *refresh mode* the stored data and display are
updated at each triggered sweep. In the *arm and release mode*
the input is captured as a single trigger point. In the *hold mode*
the display is frozen at a particular instant. The timebase is
generated by a crystal clock and so is more accurate, typically
about 0.01%, than an analogue oscilloscope which is typically
about 1–2%, and also more stable. The resolution of a digital
oscilloscope is determined by the analogue-to-digital converter.
An 8-bit ADC gives a resolution of 1 part in 256, a 12-bit
ADC 1 part in 4096. An analogue oscilloscope typically has a
resolution of about 1 mm displacement on the screen and this
is thus about 1 part in 50. To create a waveform display with a
digital oscilloscope about 10 samples are required during the
rise time of a signal. This means that the maximum rise time is
determined by the sample rate, being thus

$$\text{maximum rise time} = \frac{10}{\text{sample rate}} \qquad [4]$$

Thus to display a waveform with a rise time of, say, 100 ns
means a sample rate of 100 MHz. The sample rate of the ADC
thus limits the bandwidth of the instrument. Analogue
oscilloscopes have the advantages of being able to operate at
higher bandwidths and with higher writing speeds.

**Measurements with oscilloscopes**

The oscilloscope is a very versatile instrument which can be used to make a wide range of measurements. For example, a single channel oscilloscope can be used to display waveforms and measure voltage, current, time, frequency and rise/fall times. A two-channel instrument also enables waveforms at two points in a circuit to be easily compared and thus, for example, enables phase shifts to be determined.

For the measurement of voltage, e.g., the peak-to-peak value of an alternating waveform, it is necessary for the probe used to be properly compensated and the control for the vertical attenuator to be in the calibrated position. The signal is then connected to the $Y$ input and the peak-to-peak distance on the screen determined in terms of the divisions on the screen scale. This can then be translated into a voltage using the calibrated attenuator setting (volts/div.) and the probe attenuation.

$$\text{Voltage} = \text{scale divisions} \times \text{volts/div.}$$
$$\times \text{ probe attenuation} \qquad [5]$$

Direct and alternating currents can be measured by finding the potential difference across a known value resistance and the application of Ohm's law. Alternatively, a current probe can be used. The time between two points on a waveform can be determined using the calibrated timebase of the oscilloscope. Accuracy is about 5%. If the rise time of a signal is measured, correction has to be made for the rise time of the oscilloscope (see earlier and equation [1]).

The frequency of a signal can be determined if it is fed into the vertical input and the time determined between two points one cycle apart on the displayed waveform. The frequency is then the reciprocal of this time. The accuracy of this depends on the accuracy of the timebase and the accuracy with which the screen display can be read. Typically this results in an accuracy of about 5%. A more accurate method, however, is to use *Lissajous figures*. With this method the unknown frequency is compared with an accurately known frequency. The unknown frequency is fed into the vertical input of the oscilloscope and, with the internal time base switched off, the known frequency is fed into the horizontal input. The known frequency is then adjusted to give a stationary display on the screen. When the display is stationary there is a constant ratio between the frequencies of the two inputs. The pattern obtained on the screen depends on the ratio of the two frequencies and the phase difference between them. The frequency ratio is

$$\frac{\text{vertical input freq.}}{\text{horizontal input freq.}} = \frac{\text{number of horizontal loops}}{\text{number of vertical loops}} \qquad [6]$$

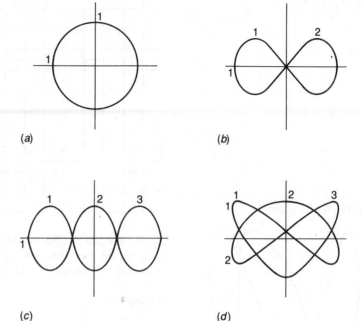

**Fig. 11.22** Lissajous figures with frequency ratios (*a*) 1-to-1, (*b*) 2-to-1, (*c*) 3-to-1, (*d*) 3-to-2

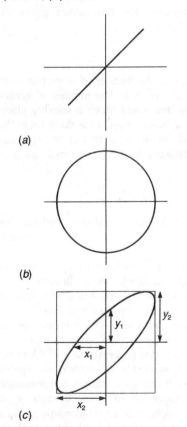

**Fig. 11.23** Phase difference: (*a*) 0°, (*b*) 90°, (*c*) in general

The number of horizontal loops is the number that could touch a suitably placed horizontal line, the number of vertical loops being the number that could touch a suitably placed vertical line. Figure 11.22 shows some examples. This method allows the unknown frequency to be determined to the accuracy of the known frequency. This can be as high as 0.001%.

Lissajous figures can also be used to determine the phase difference between two signals of the same frequency. If the phase difference between the signals is zero then the Lissajous pattern is a straight line at 45° to the horizontal axis. A 90° phase difference gives a circular pattern, assuming the two signals have the same amplitude, otherwise the pattern is an ellipse. For phase differences between 0° and 90° the pattern is an ellipse. The phase angle can be found from the ratio of the ellipse dimensions (Fig. 11.23), with

$$\sin\phi = \frac{y_1}{y_2} = \frac{x_1}{x_2}$$ [7]

**Example 3**

What is the peak-to-peak voltage and the frequency of the waveform shown in Fig. 11.24? A 10-to-1 probe is used with the vertical sensitivity set at 2 V/div. and the timebase at 10 µs/div.

*Answer*

The number of vertical scale divisions between the peaks is 5.0 and so the voltage on the *Y*-plates is 10 × 2 = 20 V. The voltage at the probe

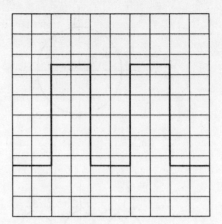

**Fig. 11.24** Example 3

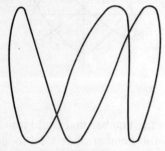

**Fig. 11.25** Example 4

tip is thus 200 V. The number of horizontal scale divisions for one cycle of the waveform is 4.0 and thus the time for one cycle is 40 μs. Hence the frequency is $1/40 \times 10^{-6} = 25$ kHz.

**Example 4**

What is the frequency ratio between the frequencies giving the Lissajous figure shown in Fig. 11.25?

*Answer*

The number of horizontal loops, i.e., the number of loops that could touch a suitably placed horizontal line, is 3. The number of vertical loops, i.e., the number of loops that could touch a suitably placed vertical line, is 1. Thus the vertical input frequency is three times that of the horizontal input. The pattern shown is in fact the type that occurs when there is a phase difference between the two frequencies of about 15°.

**Problems**

1  Describe the principles of a basic analogue cathode ray oscilloscope in terms of a block diagram indicating the main subsystems.

2  What are the characteristics required of the phosphor used for the screen of a general purpose cathode ray oscilloscope?

3  The $Y$ deflection system of an oscilloscope has a bandwidth of d.c. to 10 MHz. What would be the expected rise time?

4  Explain why probes are used with oscilloscopes instead of just connecting the signal source to the oscilloscope with wires.

5  A 10-to-1 probe has a probe head with a resistance of 9 MΩ and a capacitance of 6 pF and is connected to an oscilloscope input by a cable of capacitance 10 pF. If this gives correct compensation, what are the resistance and capacitance of the oscilloscope input?

6  A technician using a 10-to-1 probe with an oscilloscope to check the output from a square wave generator observes that the square wave has rounded corners. What could be the reason for this?

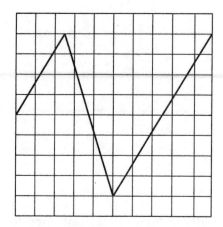

**Fig. 11.26** Problem 10

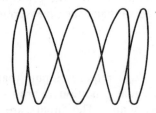

**Fig. 11.27** Problem 11

7 Suggest the forms of oscilloscope that would be suitable for making measurements in situations where the following are required:

(a) A bandwidth of the order of 100 MHz.

(b) Two signals have to be compared.

(c) Observation of the entire waveform of a slow moving signal.

(d) Observation of a rapidly repetitive waveform.

(e) Observation of non-repetitive waveform.

8 Explain what is meant by aliasing and the consequences of it in the limitations of digital storage oscilloscopes.

9 Describe the basic principles of the digital storage oscilloscope and compare its performance with that of analogue oscilloscopes.

10 Determine the peak-to-peak voltage and frequency of the signal which with a 10-to-1 probe gave the oscilloscope trace shown in Fig. 11.26. The vertical sensitivity is 1 V/div. and the timebase 1 ms/div.

11 Determine the frequency ratio between the vertical and horizontal input signals which gave the Lissajous figure shown in Fig. 11.27.

12 The following is part of the specification of a dual trace oscilloscope. Explain the significance of the information in relation to the measurements that can be made with the instrument.

*Vertical system*: two input channels
Bandwidth (−3 dB): d.c. coupled: d.c. to 15 MHz
                     a.c. coupled: 8 Hz to 15 MHz
                     rise time: 24 ns
Deflection sensitivity: 5 mV/div. to 20 V/div.,
                    in 12 calibrated steps
Voltage measurement accuracy: ±5%
Input impedance: 1 MΩ plus 40 pF (approx.)
Maximum input voltage: 500 V d.c. or a.c. peak
Operating modes: channels 1 and 2 chopped or alternate, channel 2

*Horizontal system*
Sweep speeds: 0.2 s/div. to 0.2 µs/div., in 19 calibrated steps,
                   × 5 magnifier
Maximum sweep speed: 40 ns/div.
Time measurement accuracy: ±5%, with magnifier ±7%
External input:
Bandwidth (−3 dB): d.c. coupled: d.c. to 2 MHz
                     a.c. coupled: 10 Hz to 2 MHz
Sensitivity: 1 V/div.
Input impedance: 270 kΩ plus 30 pF

*Triggering system*
Sources: internal channel 2, external, line
Trigger level: any point on the positive or negative slope
Polarity: positive or negative
Bandwidth: 15 MHz

*Display*
Phosphor: P31
Z modulation: 15 V amplitude, d.c. coupled

# 12 Counters

## Introduction

Electronic counters are digital instruments that are used to determine the number of pulses occurring in a fixed time interval, and so determine frequency, or the time interval between pulses. This chapter looks at the basic principles of such counters, how they are used to make such measurements and the errors that can occur.

## The basic counter

The basic counter/timer (Fig. 12.1) requires inputs that are square edged waves, and thus signal conditioners are used to shape the input waves to this shape. The inputs are then applied, along with a timebase signal, to a control and gating circuit. The timebase consists of a stable oscillator, usually giving a fixed frequency of 1, 5 or 10 MHz. The output from the oscillator is fed through a timebase divider which can reduce the frequency to values such as 1 kHz, 100 Hz, 10 Hz and 1 Hz. The control and gating unit can be set to operate in various modes such as frequency measurement, period measurement and time interval measurement. It has the function of permitting a controlled number of pulses from the timebase to pass through a gate to the counter. The output from the gate is passed via a counting unit to the display. The counting unit measures the number of pulses coming out of the gate. The display unit may use light emitting diodes, liquid crystals, vacuum fluorescent or cathode ray tubes to display the count number.

The accuracy of an electronic counter is almost entirely determined by the accuracy of the timebase oscillator. Generally a crystal-controlled oscillator of fixed frequency is used. This may be just a room-temperature crystal oscillator with no form of temperature compensation, a temperature-compensated crystal oscillator or an oven-controlled crystal oscillator. The uncompensated crystal oscillator is the least expensive of the three and has typically a frequency drift of

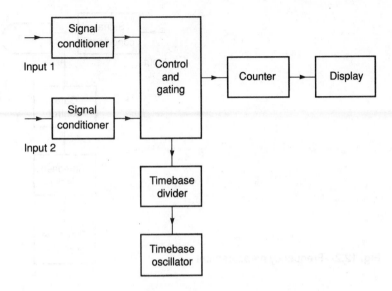

**Fig. 12.1** The basic counter

about $\pm 5$ parts per million over a temperature range 0–50 °C. The temperature-compensated oscillator uses special components and circuits to compensate at least partially for the effects of a changing temperature. The result is a frequency drift of about $\pm 2$ parts per million over a temperature range 0–50 °C. The oven-controlled oscillator is the most expensive and has its crystal thermostatically controlled to an almost constant temperature. The result is a frequency drift less than $\pm 0.2$ parts per million.

**Measurement modes**

Counters can be set to operate in different modes, according to the measurement required. The different modes involve the same basic sub-elements in the counter control and gating unit, only they are interconnected in different ways. The modes are frequency, period and time interval.

When used to measure *frequency* the arrangement of the sub-elements is as shown in Fig. 12.2. The input signal, for which the frequency is required, is made square by the signal conditioner, to give a series of square pulses with the same frequency as the input analogue signal. It is then fed into the gate. This is opened for a controlled time $t$, the time being determined by the timebase signal. The result is a count of the pulses in the squared input signal that occur in this time interval. If there are $n$ pulses then the frequency is $n/t$. Thus, if the timebase supplied a frequency of 1 kHz to the gate and opens it for just one cycle, i.e. 1 ms, then if 120 pulses were recorded during the time the gate was open, the frequency would be 120 kHz.

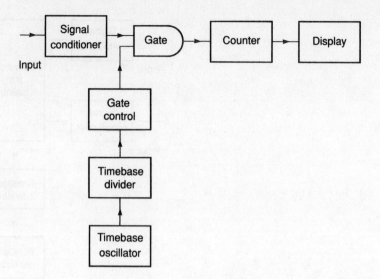

**Fig. 12.2** Frequency measurement

The measurement of low frequencies presents the problem of requiring long measurement times if there is to be reasonable resolution. For example, to measure a frequency of 10 Hz would require a measurement time of 10 s to give 100 pulses for counting and hence a resolution of 1 in 100. This problem can be overcome if the input frequency is multiplied by a constant factor of 100 or 1000 before being counted. An alternative is to use the counter to measure the time between successive pulses of the input signal; see the later discussion on the measurement of a period of time.

The speed with which the control and gating system can operate is such that the upper limit of frequency measurement is about 500 MHz. Dividing the frequency of the input by a factor of perhaps 2 or 4 enables this limit to be extended. The upper limit then becomes about 2 GHz. This does, however, reduce the resolution. An alternative, which allows measurements at higher frequencies, is to mix the input frequency with another frequency; typically some value between 10 and 500 MHz is used. The combined signal has an amplitude which varies with a frequency equal to the difference between the two frequencies. The difference between the two frequencies can then be counted. Such an arrangement is called a *heterodyne frequency converter* and permits the counter to be used for frequencies up to about 25 GHz.

This frequency mode of operation can also be used to obtain frequency ratios. For this, the timebase oscillator is disconnected and replaced by the lower frequency input signal. The other, higher frequency, input signal is thus counted for a time determined by the lower frequency signal. Thus if the lower frequency signal was 1 kHz then the gate would be

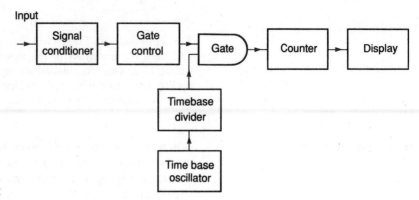

**Fig. 12.3**   Period measurement

opened for possibly one cycle, i.e. 1 ms. If the other input had a frequency of, say, 200 kHz then during that time the number of pulses counted would be 200. This is then the ratio of the two frequencies.

When used to measure a *period* of time the arrangement of the sub-elements is as shown in Fig. 12.3. The squared input is used to control the open and close operation of the gate and so determine the number of timebase pulses allowed through to the counter. The gate opens at a selected point on the input waveform and closes at the same point in the next cycle.

When used to measure a *time interval* the arrangement of the sub-elements can be of the form shown in Fig. 12.4. One input is used to open the gate and the other input to close it, so determining the number of pulses from the timebase allowed through to the counter.

**Example 1**

What is the resolution and the maximum frequency that can be displayed on a 7-digit counter if the gate time is set as 1.0 s?

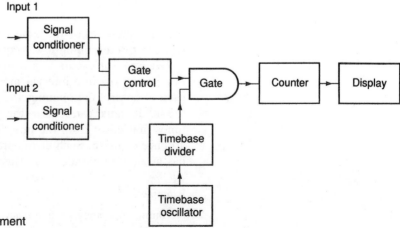

**Fig. 12.4**   Time interval measurement

*Answer*

The resolution with 7 digits will be 1 part in $10^7$. Since the gate time is 1.0 s then the frequency of the input from the timebase to the gate will be 1 Hz. Thus the smallest frequency change that can be resolved is 1 Hz. The maximum frequency that can be displayed is 9 999 999 Hz, since the least significant digit must correspond to 1 Hz.

**Measurement errors**

When the counter is used in the frequency mode there are two main sources of error: timebase errors and gating errors. When used in the period or time interval modes there are also trigger errors.

*Timebase errors* arise from changes in the frequency of the timebase oscillator. Such changes can occur as a result of calibration errors, short term changes in crystal stability and long term stability changes. The timebase oscillator is usually calibrated by using an oscilloscope to compare the frequency with that of a standard frequency transmitted by a radio station. Short term crystal stability errors are caused by momentary variations in frequency as a result of the effects of voltage transients, shock and vibration. Such effects can be minimized by taking measurements over long gate times, e.g., tens of seconds. Long time stability errors arise as a result of ageing and deterioration of the crystal, so presenting a case for calibration checks.

*Gating errors* occur because the opening and closing of the gate are not synchronized with the input signal. Thus the situation shown in Fig. 12.5 could occur, the same input signal leading to two different counts depending on when the gate was opened. Gating error gives a ±1 count error in the least significant digit of a reading. With low frequencies when only a small number of pulses are counted, this error may have a significant effect on the result; with a high frequency and a large number of counts the error is much less significant. For this reason it is preferable to use period measurements, i.e., the measurement of the time between successive pulses, as a means of determining low frequencies.

*Trigger error* occurs as a result of noise on the input signal. For period or time interval measurements the input signal is used to control the opening and shutting of the gate. Noise on the input signal can lead to the gate opening and/or closing at times that differ from the times it would have operated in the absence of noise. Such errors are minimized if the input signal has a large amplitude and a fast rise time.

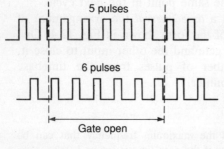

**Fig. 12.5** Gating error

**Example 2**

A counter operating in frequency mode has a gating error of ±1 count and a timebase error of 5 parts in a million. What is the percentage error when the counter is used to measure a frequency of 1 kHz?

*Answer*

The error is the sum of the two errors and so is

error = ±(1 count + 1 kHz × 5 parts per million)

= ±(1 count + 0.005 counts)

The percentage error is thus

$$\text{percentage error} = \pm \frac{1.005}{1000} \times 100\% \approx 0.1\%$$

### Example 3

A counter operating in period mode has a gating error of ±1 count. What will be the percentage error when a frequency of 50 Hz is measured when the timebase supplies a signal of 1 MHz?

*Answer*

At 50 Hz the gate is opened for 1/50 = 0.02 s. During that time the number of pulses counted will be

number = $0.02 \times 10^6$ = 20 000

The gating error means an error of 1 count in 20 000. Thus the frequency error is

$$\text{error} = \pm \frac{1}{20\,000} \times 50 = \pm 0.0025\,\text{Hz}$$

The percentage error is thus

$$\text{percentage error} = \pm \frac{0.0025}{50} \times 100\% = \pm 0.05\%$$

**Problems**

1 What is the frequency of an input signal to a counter in frequency measurement mode if the gate is opened for 100 ms and 2354 pulses are counted?

2 What is the resolution of a counter with a $3\frac{1}{2}$ decade display?

3 In the determination of low frequencies, why is the more accurate procedure to use the period mode rather than frequency mode?

4 Explain how the basic counter can be modified to enable (*a*) low frequencies and (*b*) high frequencies to be measured.

5 What is the resolution and maximum frequency that can be displayed on the 6-digit display of a counter in frequency mode if the gate time is 0.1 s?

6 A counter operating in frequency mode has a gating error of ±1 count and a timebase error of 3 parts per million. What will be the percentage error when a frequency of 2 MHz is measured?

7 A counter operating in period mode has a gating error of ±1 count. What will be the percentage error when a frequency of

1 kHz is measured when the timebase supplies a signal of 1 MHz?

8   A counter includes the following information as part of its specification. Explain the significance of the information.

*Frequency mode*
Range: d.c. to 50 MHz (direct)
Gate times: manual: 1 ms to 100 s in decade steps
automatic: gate times up to 1 s selected automatically to avoid overspill

*Period mode*
Range: 1 μs to 1 s
Time unit: 1 μs

*Time interval mode*
Input channel: double line, start channel 2, stop channel 1
Time range: 100 ns to $10^4$ s

# 13 Signal sources

Instruments that generate signals are widely used to provide test signals that can evaluate the performance of electronic systems. There are various types of signal sources, depending on the type of waveform required and the frequency range. The term *oscillator* is usually used for a signal source that generates only sinusoidal waveforms, the term signal generator for one that not only generates sinusoidal waveforms but also can be used to modulate a signal. The term *function generator* is used for an instrument able to provide a variety of different wave shapes, such as sinusoidal, square, pulse and triangular. Synthesizers are oscillators in which the signal is built up digitally.

**Output impedance**

An important consideration with any signal source is the output impedance. A source can be considered to be an ideal voltage source with no internal impedance in series with an impedance, this being the so-called output impedance. When the source is connected to a circuit, maximum power is dissipated in the circuit only when the output impedance equals the impedance of the circuit. Thus it is desirable to match the output impedance and the circuit impedance. Signal sources are usually designed to have output impedances of either 600 or 50 $\Omega$. Audio frequency circuits, such as telephone circuits, are designed to have a characteristic impedance of 600 $\Omega$, while coaxial cables used with radio frequency signals are designed to have a characteristic impedance of 50 $\Omega$. The characteristic impedance is the impedance which, if connected across the input to the circuit, also appears across its output and so means matching occurs when the output impedance of the signal source equals the characteristic impedance.

**Oscillators**

Oscillators are instruments that generate sinusoidal signals, the

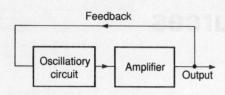

**Fig. 13.1**   Basic oscillator

term low frequency oscillator being used for the frequency range from about 1 Hz to 1 MHz and high frequency oscillator for 100 kHz to 500 MHz. The term signal generator is often used for oscillators that can also be used to modulate a signal. All oscillators contain three basic elements: an oscillatory circuit which determines the frequency, an amplifier to provide amplification and a usable output signal, and a feedback circuit to divert some of the amplified output back to the input to compensate for losses in the oscillatory circuit and maintain the oscillations (Fig. 13.1). There are two main types of oscillatory circuit that are used for oscillators: inductance–capacitance ($LC$) and resistance–capacitance ($RC$).

An inductance–capacitance circuit resonates at a frequency $f$ given by

$$f = \frac{1}{2\pi\sqrt{LC}} \tag{1}$$

The most widely used oscillatory circuits which are based on the $LC$ resonance are the *Hartley* and the *Colpitts*. Such oscillators are used to provide frequencies in the range 10 kHz to 100 MHz. They are not suitable for lower frequencies because such resonant frequencies would require large inductances and capacitances.

$RC$ oscillator circuits are circuits which are frequency selective. The two most common forms of $RC$ oscillatory circuit are the *Wien bridge* and the *phase shift oscillator*. The Wien bridge was discussed in Chapter 9. When the bridge is balanced (equations [35] and [36] in Chapter 9 with the components as in Fig. 9.21):

$$C_4 = \frac{(R_1/R_2)C_3}{1 + \omega^2 R_3^2 C_3^2}$$

$$R_4 = \frac{R_2(1 + \omega^2 R_3^2 C_3^2)}{\omega^2 R_3 R_1 C_3^2}$$

Multiplying these two equations gives

$$C_4 R_4 = \frac{1}{\omega^2 R_3 C_3}$$

With $R_1 = R_2 = R$ and $C_3 = C_4 = C$ then, since $\omega = 2\pi f$,

$$f = \frac{1}{2\pi\sqrt{RC}} \tag{2}$$

The Wien bridge oscillator is widely used for the frequency range 20 Hz to 20 kHz and has an upper limit of 1 MHz. Phase shift oscillators are another form of $RC$ oscillator, having the advantage over the Wien bridge of giving a wide frequency range, up to 10 MHz, but the disadvantage of not being as frequency stable.

Important parameters that have to be considered in selecting an oscillator are the frequency range, the frequency stability, the accuracy with which the frequency is indicated by the dial/display on the instrument, the dial/display resolution, the power/voltage output, the stability of the output (i.e., the ability to maintain a constant voltage amplitude output), the waveform distortion and the output impedance.

A distorted sine wave can be considered to be a pure sine wave to which a number of harmonics have been added (see Chapter 14), such distortion being referred to as harmonic distortion. The amount of distortion can then be specified in terms of the percentage of the r.m.s. voltage which is due to harmonics. A typical oscillator might be specified as less than 2%. Alternatively the distortion might be specified as the amount of power in the harmonics relative to that in the pure sine wave. A typical value might be −40 dB.

The output impedance of oscillators is designed to remain constant over the frequency range of the instrument. With audio frequency oscillators it is usually 600 Ω while with radio frequency oscillators it is 50 Ω. These values are chosen because the characteristic impedance of audio frequency systems has been standardized at 600 Ω while for radio frequencies the coaxial cable used usually has a characteristic impedance of 50 Ω. It is necessary for the impedance of the oscillator to be matched to that of the load across its terminals if there is to be maximum power transfer.

**Pulse and square wave generators**

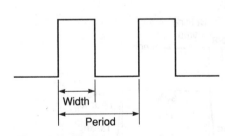

Width
Period

**Fig. 13.2** Pulses

A pulse generator is one that produces pulses for which the width of the pulse remains constant but the time between pulses is varied (Fig. 13.2). A square wave generator is one that produces pulses with a width equal to half the cycle time, the width changing to maintain this relationship as the cycle time is varied. The term *duty cycle* is used to define the ratio of the pulse width to pulse cycle time or period:

$$\% \text{ duty cycle} = \frac{\text{pulse width}}{\text{pulse period}} \times 100\% \qquad [3]$$

Thus a square wave has a duty cycle of 50%. Circuits to produce pulses or square waves may be pulse-shaping ones that convert a sine wave output from an oscillator into a pulse train (Fig. 13.3) or a circuit that directly produces pulses, e.g., a multi-vibrator. Some pulse generators allow manual triggering, a single pulse being produced when a button is depressed on the instrument front control panel. A double pulse mode allows two pulses to be produced with a delay between them. In the burst mode pulses are produced during a

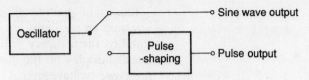

**Fig. 13.3** Sine and pulse/square wave generator

controlled time. The output impedance of pulse generators is typically $50\,\Omega$.

There are a number of terms used to describe the characteristics of the pulses produced by a pulse generator (Fig. 13.4).

1 The *amplitude* of a pulse is measured from the level at which the pulse starts, i.e., the base line, to the steady state pulse value.

2 The *base line offset* is the amount by which the base line is displaced from the 0 V line.

3 The *pulse rise time* is the time required for a pulse to rise from 10% to 90% of its normal amplitude.

4 The *pulse fall time* is the time required for a pulse to fall from 90% to 10% of its normal amplitude.

5 The *pulse width* is the time the pulse takes from the 50% amplitude point on the leading edge to the 50% point on the trailing edge.

6 The pulse *preshoot* is the deviation of the base line at the start of the pulse.

7 The pulse *undershoot* is the deviation of the base line immediately following the end of the pulse.

8 The pulse *overshoot* is the deviation from the peak value of the pulse immediately following a rising edge.

9 *Ringing* is the oscillation of the pulse height immediately following a rising edge.

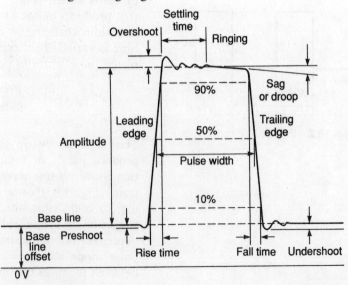

**Fig. 13.4** Pulse parameters

10 The *settling time* is the time required for the ringing to decrease to a given percentage, usually 1–5%, of the overshoot.

11 *Droop* or *sag* is the amount by which the peak value of the pulse decreases during the pulse.

12 The *pulse repetition rate* is the frequency with which a pulse occurs and is equal to the reciprocal of the pulse cycle time or period.

**Example 1**

A pulse generator is required to produce pulses with a duty cycle of 20% and a period of 1 ms. What is the required pulse width?

*Answer*

Duty cycle is given by equation [3] as

$$\% \text{ duty cycle} = \frac{\text{pulse width}}{\text{pulse period}} \times 100\%$$

Hence the pulse width is $0.20 \times 1 = 0.20$ ms.

**Sweep frequency generator**

The sweep frequency generator is a sine wave generator for which the frequency can be electronically swept over a range of frequencies. For example, the generator may sweep from 1 MHz to 1.5 MHz in 10 kHz steps. The sweep generator is particularly useful for determining the frequency response of items such as amplifiers and filters. Figure 13.5 shows the basic test arrangement. The sweep time base supplies a sweep through a frequency band to the test item and the output, after being rectified and filtered, is fed to the vertical input of an oscilloscope. The sweep timebase also supplies a signal to the horizontal displacement input of the oscilloscope. The result is a display which shows, for the frequency band used, how the amplitude of the output from the test item varies with frequency.

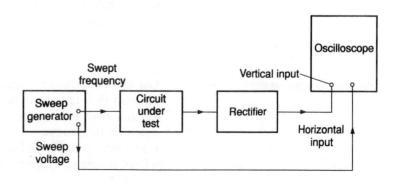

**Fig. 13.5** Test arrangement for frequency response

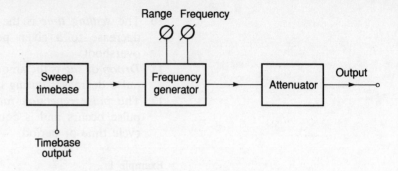

**Fig. 13.6** Basic sweep frequency generator

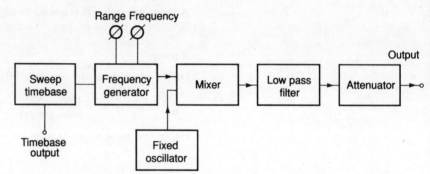

**Fig. 13.7** Heterodyne controlled sweep frequency generator

The basic sweep frequency generator (Fig. 13.6) consists of an adjustable timebase which sweeps the frequency of a frequency generator through a specified range of frequencies. The timebase can typically be adjusted to give sweeps over time intervals of between 10 ms and 100 s. There are two modes for setting the frequency range. The stop–start or $f_1 - f_2$ mode allows the stop and start frequencies to be specified. The delta frequency ($\Delta f$) mode allows the central frequency of the sweep to be specified and the difference in frequency on either side of this central frequency over which the sweep is to be made. The frequency bands within which sweep generators operate are typically 0.001 Hz to 100 kHz, 100 kHz to 1500 MHz and 1 GHz to 200 GHz. Wider band sweeps can involve manually or electronically switching between generators. An alternative is the heterodyne controlled sweep frequency generator (Fig. 13.7). In this method the sweep frequencies are mixed with the frequency from a fixed oscillator.

**Frequency synthesizer**

Oscillators produce frequencies by adjustment of some element in a tuning circuit and can be tuned in a continuous manner over a frequency range. An alternative, which gives much better accuracy and stability of frequency, is the frequency synthesizer. The synthesizer derives its output from a fixed frequency, highly stable oscillator and covers a range of

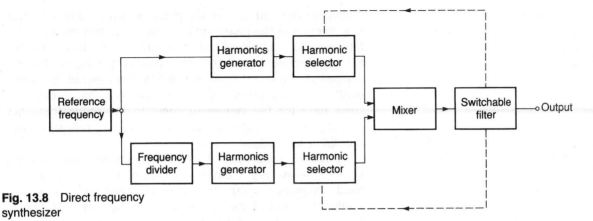

**Fig. 13.8** Direct frequency synthesizer

frequencies in a number of steps rather than continuously. There are two basic types of synthesizer, direct and indirect.

Figure 13.8 shows the basic form of the direct synthesizer. A stable crystal oscillator supplies a fixed frequency which drives one harmonics generator to produce a signal containing many harmonics. Also the oscillator, after frequency division, supplies a second harmonics generator. Harmonics can be selected from each generator and mixed. Since the mixing produces an output of the two input frequencies, the sum of the two frequencies and the difference of the two frequencies, a tunable filter is used to pass only the sum or the difference frequency. By switching the harmonics selected at the harmonic selectors so the output frequency can be switched in a number of steps.

Figure 13.9 shows the basic form of the indirect synthesizer, this using a phase-locked loop to produce an output which is an integral multiple of the frequency of a constant frequency crystal oscillator. A voltage-controlled oscillator is used to produce an output which is tuned electrically by using a variable voltage. The output from the oscillator is fed, after passing through a divider, to a phase detector where its phase is compared with that from a constant frequency crystal

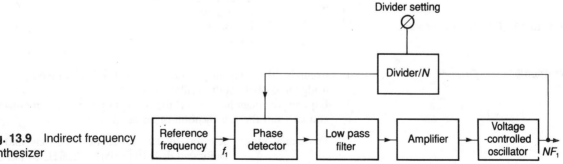

**Fig. 13.9** Indirect frequency synthesizer

oscillator. The output from the phase detector is a signal which is a function of the phase angle between the two inputs. After filtering and amplification, this output is used to control the voltage-controlled oscillator. The result is that the oscillator is controlled to give an output which is N times that of the crystal oscillator. A disadvantage of this system is that it can take some time for the system to come to the stable frequency output, also the output is likely to show more distortion than the direct method.

By combining a number of frequency synthesizer circuits, with each covering a different range of frequencies, a wide band frequency synthesizer can be produced. A typical synthesizer may have a frequency range of 100 kHz to 1000 MHz in steps of 100 Hz. Synthesizers are also available for higher frequency ranges, e.g., 0.01 to 30 GHz in steps of 1 kHz. The signals are also generally available with either frequency or amplitude modulation.

**Function generators**

Function generators are able to produce several different waveforms from the same instrument, the three basic forms being triangular, sinusoidal and square. The frequency range is typically 1 Hz or less to 20–50 MHz. Figure 13.10 shows the basic form of such an instrument. The output from a square wave oscillator is integrated to give a triangular waveform. The use of a suitable shaper can convert this into a sinusoidal waveform. Thus outputs of square, triangular and sinusoidal waveforms are possible. Other shaper circuits are often available to give pulse, ramp and sawtooth waveforms.

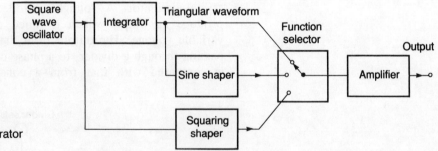

**Fig. 13.10** Function generator

**Problems**

1. Compare the frequency ranges of the Hartley, Colpitts, Wien bridge and phase shift oscillators.
2. Explain the significance of the information in the following extract from the specification of an oscillator:

    Frequency range: selected by five range push buttons and a calibrated dial, 10 to 100 Hz, 100 Hz to 1 kHz, 1 to 10 kHz, 10 to 100 kHz, 100 kHz to 1 MHz

Accuracy: ±3% of full-scale range
Output amplitude: continuously variable 0–2.5 V r.m.s. for both
　　　　　　　sine and square waves
Sine wave distortion: Better than 2% up to 100 kHz
　　　　　　　　　　Typically <0.5% over 50 Hz–10 kHz
Square wave rise time: <100 ns, typically 50 ns
Output impedance: 600 Ω

3　A pulse generator produces pulses of width 20 ms and period 80 ms. What is its duty cycle?

4　A pulse generator is said to have rise times and fall times that can be set to within the range 0.5 ns to seconds. Explain the significance of the data.

5　A pulse generator can have its duty cycle set to within 10 to 90%. Explain what this means.

6　Explain how a sweep frequency generator can be used to determine the frequency characteristics of an amplifier.

7　Explain the difference in mode of operation of direct and indirect frequency synthesizers.

# 14 Signal analysis

Many electrical waveforms encountered in electrical and electronic equipment and circuits, e.g., a square waveform, are not sinusoidal. However, all repetitive electrical waveforms can be considered to be made up of a combination of sinusoidal waveforms of various frequencies and amplitudes. This is the basis of what is called *Fourier's theorem*. Thus, for example, a square waveform can be considered to be made up from a combination of sinusoidal waveforms. This chapter is thus concerned with the analysis of signals to determine the frequencies of the sinusoidal constituent components. Consequently it is also concerned with distortion measurement where the distortion can be considered to be due to the addition of harmonics to a sinusoidal signal. A brief consideration of the measurement of noise is also included.

## Fourier's theorem

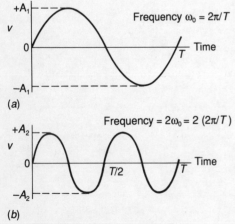

(a)

(b)

**Fig. 14.1** Sinusoidal signals:
(a) amplitude $A_1$, frequency $\omega_0$;
(b) amplitude $A_2$, frequency $2\omega_0$

A periodic signal is one that repeats itself at regular intervals, the time between successive repetitions being called the *periodic time T*. A sinusoidal signal (Fig. 14.1) can be represented by

$$v = A_1 \sin \omega_0 t$$

where $A_1$ is the amplitude of the signal and $\omega_0$, the angular frequency, is $2\pi/T$. A signal with twice the frequency, and a different amplitude $A_2$, may be represented by

$$v = A_2 \sin 2\omega_0 t$$

A signal with three times the frequency, and a different amplitude $A_3$, may be represented by

$$v = A_3 \sin 3\omega_0 t$$

The above equations all describe sinusoidal signals that have started off with $v = 0$ at $t = 0$. When this is not the case we can represent the signals by

$$v = A_1 \sin(\omega_0 t + \phi_1)$$
$$v = A_2 \sin(2\omega_0 t + \phi_2)$$
$$v = A_3 \sin(3\omega_0 t + \phi_3)$$

where $\phi_1$, $\phi_2$, and $\phi_3$ are the phase angles with respect to $t = 0$.

According to *Fourier's theorem* we can consider any periodic signal to be made up of a combination of sinusoidal waves. Thus we can consider a periodic signal to be presented by

$$v = A_1 \sin(\omega_0 t + \phi_1) + A_2 \sin(2\omega_0 t + \phi_2)$$
$$+ A_3 \sin(3\omega_0 t + \phi_3) + \ldots + A_n \sin(n\omega_0 t + \phi_n) \, [1]$$

If the signal includes a d.c. component $A_0$ then

$$v = A_0 + A_1 \sin(\omega_0 t + \phi_1) + A_2 \sin(2\omega_0 t + \phi_2)$$
$$+ A_3 \sin(3\omega_0 t + \phi_3) + \ldots + A_n \sin(n\omega_0 t + \phi_n) \, [2]$$

where $\omega_0 = 2\pi/T$ and is called the *fundamental frequency* or *first harmonic*. The frequency $2\omega_0$ is called the *second harmonic*, $3\omega_0$ the third harmonic and so on to $n\omega_0$ as the $n$th harmonic. $A_1$, $A_2$, etc. are the amplitudes of the various harmonics.

Figure 14.2 shows how an almost square waveform can be built up with the fundamental and the third harmonic, the amplitude of the third harmonic being one-third that of the fundamental. For such a wave we can write

$$v = A_1 \sin\omega_0 t + (1/3)A_1 \sin 3\omega_0 t$$

A better approximation to a square waveform is given by adding higher odd harmonics,

$$v = A_1 \sin\omega_0 t + (1/3)A_1 \sin 3\omega_0 t + (1/5)A_1 \sin 5\omega_0 t$$
$$+ (1/7)A_1 \sin 7\omega_0 t + \ldots$$

Figure 14.3 shows a graph of the amplitude of the constituent sinusoidal waves plotted against the frequency for a square waveform, i.e., the waveform giving the above equation. Such a graph is referred to as the *amplitude spectrum* and shows that the square waveform is made up of frequencies at $\omega_0$, $3\omega_0$, $5\omega_0$ and $7\omega_0$ and indicates their amplitudes. A graph of the phases of each of the constituent sinusoidal waves plotted against the frequency gives what is called the *phase spectrum*. Both graphs together constitute what is called the *frequency spectrum*.

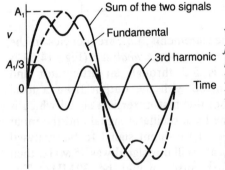

**Fig. 14.2** $\quad v = A_1 \sin\omega_0 t + \tfrac{1}{3}A_1 \sin 3\omega_0 t$

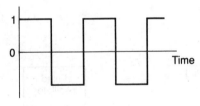

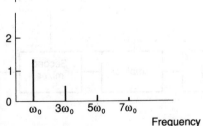

**Fig. 14.3**  Amplitude spectrum for a square wave

## Wave analyser

The wave analyser is an instrument used to measure the amplitude of a single frequency in a complex waveform. Essentially the instrument is a *frequency selective voltmeter* which is tuned to a frequency and then measures the amplitude at that frequency. For measurements in the audio range the instrument has a narrow pass band filter which can be tuned to the required frequency. The output from the filter is amplified and displayed on a meter, the meter reading then being the amplitude of the frequency selected by the tuning of the filter. Figure 14.4 shows the basic form of the instrument. The bandwidth of the instrument is typically about 1% of the selected frequency.

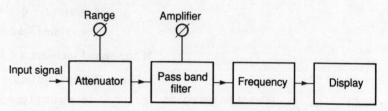

**Fig. 14.4**  Audio frequency range wave analyser

For radio frequency measurements, i.e., frequencies in the megahertz range, a *heterodyne wave analyser* (Fig. 14.5) is used. The input signal is fed through an attenuator and amplifier before being mixed with the output from a local oscillator. The output from this first mixer is a signal which is the difference between the local oscillator signal and the input signal. Thus, for example, if the input signal to be analysed was at 15 MHz and the local oscillator signal was 45 MHz, then the output from the first mixer would be 30 MHz. The frequency of the local oscillator is adjusted so that the output

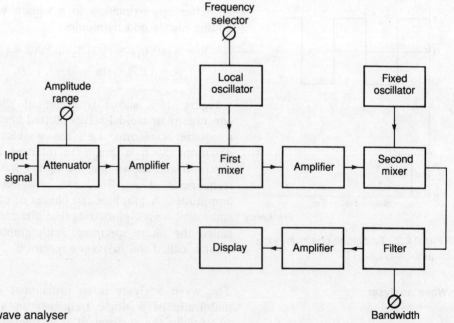

**Fig. 14.5**  Heterodyne wave analyser

from the first mixer gives a constant frequency which is in the pass band of the amplifier forming the next stage of the instrument. Adjustment of the local oscillator frequency thus enables the input signal frequency to be selected. The output from the amplifier is then mixed with the output from a fixed frequency oscillator. This frequency is fixed so that the output from the second mixer is centred on zero frequency. Thus if the output from the first mixer was 30 MHz then the fixed frequency oscillator would be at 30 MHz. This output passes through a filter with a controllable bandwidth, before final amplification and display on a meter.

**Spectrum analysers**

Spectrum analysers are instruments that resolve an electrical waveform into its constituent sinusoidal waveforms and then display them as the amplitude spectrum. Spectrum analysers can be considered to fall into two main groups: those using band pass filters and those that convert the waveform into digital signals. The form of spectrum analyser which converts the waveform into digital signals is called a Fourier analyser and is dealt with later in this chapter. Band pass filters are electrical filters which only allow a defined band of signal frequencies to pass, all other frequencies being heavily attenuated. There are two basic forms of spectrum analyser using such filters, one using a bank of band pass filters with each one passing a different frequency and the other where a single tunable filter is used and the frequency it passes is varied as it sweeps across the frequency range being investigated.

With the *parallel bank of filters*, the frequency range being investigated is broken down into a number of frequency bands, each filter being designed to respond to just one particular frequency band. A spectrum analyser for audio frequencies might have 32 such band pass filters. The output from each filter is scanned in sequence and produces a vertical deflection on a cathode ray tube, the scan generator simultaneously producing a horizontal deflection (Fig. 14.6). The result is that the screen of the cathode ray tube displays the amplitude spectrum. For such an analyser to be able to distinguish between frequencies relatively close together, i.e., to have what is termed high *resolution*, it is necessary for each filter to have a narrow pass band. But the narrower the pass band, the more filters are required to cover the frequency spectrum. Thus a high resolution analyser of this form can be quite expensive and is generally only made for a limited frequency range, e.g., the audio range from 0 Hz to about 20 kHz. For a wide band with narrow resolution the swept frequency method is generally used, since it requires just one filter.

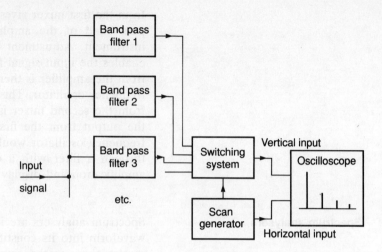

**Fig. 14.6** Parallel filter bank analyser

The basic principle of the *swept frequency analyser* is that just one band pass filter is used. The frequency it transmits can be varied and is swept over the frequency range being investigated, the size of the signal at each position being monitored. High resolution is possible since the bandwidth of the filter can be made quite narrow. The easiest way to sweep the frequency which can pass through the band pass filter is to use a fixed frequency band pass filter and cause the input frequency to be altered in a controlled way so that it sweeps through the filter frequency. Figure 14.7 shows how such a system is generally realized in practice. The input signal is mixed with the output from a voltage-controlled local oscillator, the input signal being said to be heterodyned. The output from the mixer is a signal which has frequencies equal to the difference between the input signal frequencies and the local oscillator frequency. A band pass filter is then used to isolate a frequency band from the 'difference' signal. The output from the filter, after amplification, is then fed to the vertical input of a cathode ray tube. The local oscillator frequency is swept so that the 'difference' frequency is varied, with the result that when the difference corresponds to the pass band of the filter there is an output to the cathode ray tube. The sweep generator used to control the sweep of the local oscillator also controls the horizontal deflection sweep of the cathode ray tube. Thus the result is that the cathode ray tube displays the amplitude spectrum.

The above represents the basic principle of the swept frequency spectrum analyser. In practice the instrument tends to be more of the form shown in Fig. 14.8. The example given is for a spectrum analyser which analyses signals up to 1.25 GHz. The input first passes through an attenuator and low pass (LP) filter. The low pass filter is used to restrict the upper

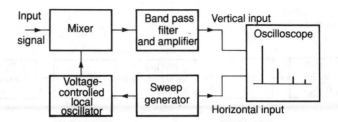

**Fig. 14.7**  Swept frequency analyser

limit of the range of frequencies to be analysed. This is necessary since the sweep frequency range from the voltage-controlled local oscillator is chosen to ensure that the 'difference' frequencies emerging from the first mixer are all at a higher frequency than the input and so can be isolated from the input frequencies.

The first local oscillator might thus have a sweep frequency from 2.0 to 3.3 GHz, so giving an output 'difference' signal which can vary from 2.0 to 3.3 GHz, for a 0 Hz signal, and $(2.0 - 1.25) = 0.75$ GHz to $(3.3 - 1.25) = 2.05$ GHz for a 1.25 GHz signal. Only when the 'difference' signal is 2.05 GHz can the output be transmitted by the band pass filter/amplifier which is the next stage. This signal is then mixed with a lower frequency signal provided by another local oscillator, e.g., 1.5 GHz, to give a 'difference' signal of 0.55 GHz. This then passes through a second band pass filter/amplifier to yet another mixer. Here it is mixed with a lower frequency signal, e.g., 0.50 GHz, to give a 'difference' signal of 50 MHz. Following another band pass filter/amplifier this might be yet further reduced in frequency by mixing with a 47 MHz frequency to give a 'difference' signal of 3 MHz. A low final frequency is desirable because narrow band pass filters, i.e., high resolutions, are more easily achieved at low frequencies and the detector that can be used is cheaper. The detector gives an output related to the amplitude of the 3 MHz signal. The detector can be set to give an output which is directly proportional to the amplitude, i.e., a linear scale, or proportional to the log of the amplitude. This means that the vertical deflection scale on the cathode ray tube can be calibrated to give readings in decibels, e.g., each vertical scale division corresponding to perhaps a 10 dB difference in amplitude. The last unit before input to the vertical deflection plates of the cathode ray tube is a video filter. This is a low pass filter and is used to reduce the noise level of the signal. The range of the instrument may be changed to a lower frequency by sweeping, say, the 0.50 GHz oscillator instead of the 2.0–3.3 GHz oscillator.

Swept frequency analysers are available for use with frequencies from about 5 Hz to 220 GHz. Generally, in use the procedure is to set the central vertical line of the scale, often

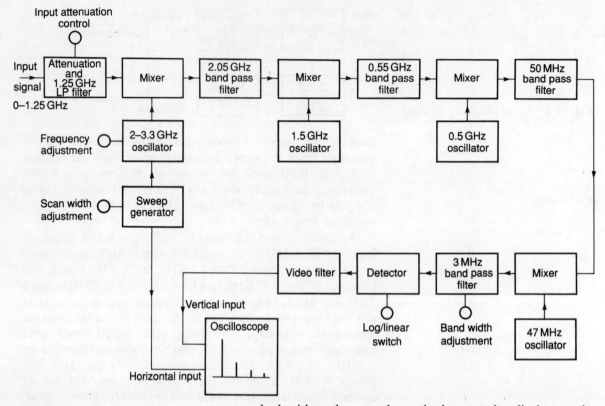

**Fig. 14.8** A 1.25 GHz swept frequency analyser

marked with a dot, on the cathode ray tube display to the central frequency required for the swept range and then set the range of frequencies to be displayed on the display by setting the frequencies to be scanned for each horizontal scale division. The bandwidth of the band pass filter varies from about 1 Hz to 3 MHz, depending on the frequency range over which the spectrum is being monitored. The bandwidth can be adjusted and determines the resolution of the instrument; the smaller the bandwidth, the greater the resolution, i.e., the closer can be the frequencies of two signals and they can be still discerned as being two. The rate at which the frequency spectrum can be swept depends on the bandwidth of the filter; the smaller the bandwidth, the greater the sweep time. This is because the smaller the bandwidth, the greater number of points there are within the frequency range at which amplitude measurements are made. The faster the sweep rate, the more rapidly a change in spectrum will be discerned. An important factor in determining the usefulness of an analyser is the *dynamic range*. This is the range of signals between the smallest that can be detected above the noise of the system and the largest signal that does not cause any spurious signals greater than the smallest signal that can be detected. Typically, this varies from about −130 dBm to +30 dBm (the dBm unit of power is the power level in decibels when compared with a power level of 1 mW; thus for example 100 mW is 10 lg (100/1) = 20 dBm). An important part of the specification of a spectrum analyser is the stability. Long term stability is usually specified in terms of the frequency drift within some specified time, e.g., Hz/hour.

The type of analyser to be used depends on the frequency spectrum being monitored, the analyser involving a parallel bank of filters or the Fourier analyser (see later this chapter) being used for low frequencies, up to about 100 kHz, and the swept frequency analyser, using heterodyning, for higher frequencies. Analysers are used to carry out such measurements as the analysis of the signals from oscillators, the output of modulators, the frequency response of components and noise measurements.

**Example 1**

A spectrum analyser gave as a display on the cathode ray tube just a single spectral line, as illustrated in Fig. 14.9, regardless of the bandwidth of the band pass filter. What was the waveform of the input to the analyser?

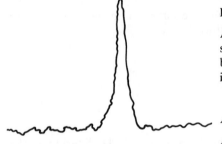

**Fig. 14.9** Example 1

*Answer*

A single spectral line, regardless of the bandwidth of the filter, means that there was just one frequency in the input. This can only be the case if the input was a pure sinusoidal signal. The bandwidth of the filter is an important point in determining whether the signal is a pure sinusoidal signal, since if the bandwidth was too wide this might mean that there were more than frequency present, which had not been resolved.

**Example 2**

Figure 14.10 shows the display obtained with an amplitude-modulated sinusoidal signal. The central vertical line of the display has been set to a frequency of 100 kHz and the frequency scan at 4 kHz per horizontal scale division. What are the frequencies of the carrier wave and its sidebands?

*Answer*

The carrier wave is on the central vertical line and so has a frequency of 100 kHz. The two sidebands are one scale division displaced from this and so are at 96 and 104 kHz. This would indicate that the carrier wave has been modulated by a 4 kHz signal since the frequency of the sidebands $f_{SB}$ are given by

$$f_{SB} = f_C \pm f_A \qquad [3]$$

where $f_C$ is the frequency of the carrier and $f_A$ the frequency of the amplitude modulating signal.

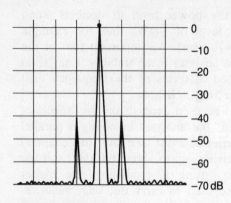

**Fig. 14.10** Example 2

**Example 3**

A swept frequency analyser has as its input an amplitude modulated signal and gives the display shown in Fig. 14.10. For an amplitude-modulated signal the modulation factor $m$ is given by

$$\frac{4}{m^2} = \frac{\text{carrier power}}{\text{power in one sideband}} \qquad [4]$$

Estimate the modulation factor for the display in Fig. 14.10.

*Answer*

Note that the spectrum analyser display is giving a vertical log scale in decibels. Taking logarithms of the above equation gives

$$\lg(4/m^2) = \lg(\text{carrier power}) - \lg(\text{sideband power})$$

Thus

$$10\lg(4/m^2) = \text{carrier power in dB} - \text{sideband power in dB}$$

For the display in the figure, the difference in the powers of the carrier and a sideband is 40 dB. Hence

$$10\lg(4/m^2) = 40$$

$$(4/m^2) = 10^4$$

Hence $m = 0.02$.

**Fourier transform analyser**

The Fourier transform analyser (Fig. 14.11) determines the magnitude and phase of each frequency component of the input to the instrument. The input signal first passes through a low pass filter to remove any out-of-band frequency components. The signal is then sampled, by the sample and hold circuit, and converted to digital form at regular intervals by the analogue-to-digital converter. The digital samples are stored in the memory until enough samples have been obtained to give a big enough signal. This is referred to as the time record. The processor then examines the signal using the mathematical principles of the Fourier transform to determine the amplitudes and phases of the frequency components. The result is then stored in the frequency memory and displayed on an oscilloscope. The Fourier transform analyser has a frequency range from d.c. to about 100 kHz and is faster at lower frequencies than the swept spectrum analyser.

**Distortion analyser**

Ideally the output from an oscillator should be a sinusoidal signal. When a sinusoidal signal is applied to an amplifier the output should be sinusoidal. In practice, however, the output signals are not perfectly sinusoidal but show some distortion. A common form of distortion is *harmonic distortion* due to non-linear behaviour of circuit elements. A consequence of

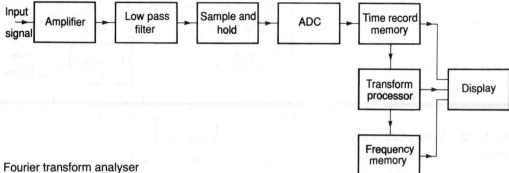

**Fig. 14.11**  Fourier transform analyser

this is that harmonics of the fundamental frequency are produced and these, when combined with the fundamental, result in a distorted output. A measure of the distortion produced by a particular harmonic is the ratio of the amplitude of that harmonic to that of the fundamental frequency.

$$\%N\text{th harmonic distortion } D_N = \frac{V_N}{V_f} \times 100\% \qquad [5]$$

where $V_N$ is the r.m.s. voltage of the $N$th harmonic and $V_f$ the r.m.s. voltage of the fundamental. The *total harmonic distortion* (THD) is defined as being

$$\text{THD} = \surd(D_2^2 + D_3^2 + \ldots + D_N^2) \qquad [6]$$

where $D_2$ is the percentage 2nd harmonic distortion, $D_3$ the percentage 3rd harmonic distortion, etc.

Measurements with a spectrum analyser on a waveform can give the amplitudes of the harmonics and the fundamental and so enable the total harmonic distortion to be calculated from the distortion values for each harmonic. Rather than determine the harmonic components of a waveform a distortion analyser can be used to determine the total harmonic distortion. In such cases, a modified version of the above definition for distortion is often used, the distortion being measured in terms of the ratio of the harmonic plus noise to the total waveform, i.e., fundamental plus harmonics plus noise. Since the harmonics are generally only a very small element in the total waveform this approximation does not lead to significant deviations in results from when the distortion is just the ratio of the harmonic to fundamental. The error is no more than 0.5% when the total harmonic distortion is less than 10%.

$$\text{THD} = \frac{\surd[(\text{harmonics})^2 + (\text{noise})^2]}{\surd[(\text{fundamental})^2 + (\text{harmonic})^2 + (\text{noise})^2]} \qquad [7]$$

Figure 14.12 shows the basic form of a *harmonic distortion analyser*. It consists of a sinusoidal signal source which can be

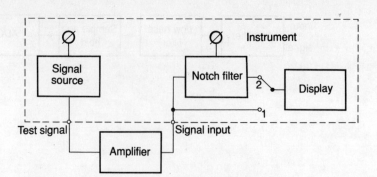

**Fig. 14.12** Harmonic distortion analyser

applied as input to an amplifier or other external component under test. The output from the amplifier is then fed back into the instrument. With the switch in position 1 the output from the amplifier passes directly to the display and the total root-mean-square amplitude $V_T$ due to the fundamental plus all the introduced harmonics plus noise is displayed. With the switch in position 2 a notch filter is introduced. A common form of notch filter is a Wien bridge (see Chapter 9). This filter removes the fundamental frequency. The display thus now indicates the root-mean-square amplitude $V_H$ due to just the harmonics and the noise. The total harmonic distortion is then given by

$$\text{THD} = \frac{V_H}{V_T} \times 100\% \qquad [8]$$

**Example 4**

A spectrum analyser indicates that the output from an amplifier contains the following signals: fundamental $V_1 = 1.0\,\text{V}$, second harmonic $V_2 = 0.02\,\text{V}$, third harmonic $V_3 = 0.005\,\text{V}$. What is (a) the percentage harmonic distortion of the second harmonic and (b) the percentage total harmonic distortion?

*Answer*

(a) Using equation [5], the percentage distortion due to the second harmonic is $(0.02/1.0) \times 100\% = 2\%$.

(b) The distortion due to the third harmonic is $(0.005/1.0) \times 100\% = 0.5\%$. Thus, using equation [6], the total harmonic distortion is

$$\text{THD} = \sqrt{(2^2 + 0.5^2)} = 2.1\%$$

**Noise measurement**

Noise is a problem that occurs with all electronic systems. There is noise generated within the system and also present in the input to the system. Often the noise generated within the system is virtually the entire noise. A standardized way of specifying the noise generated within a system is the noise factor. The *noise factor* of a device is defined as (see Chapter 3)

$$\text{noise factor } F = \frac{(\text{signal/noise ratio})_{\text{in}}}{(\text{signal/noise ratio})_{\text{out}}} \qquad [9]$$

when the system is at a standard temperature of 290 K. When expressed in decibels it is often referred to as the *noise figure*:

$$\text{noise figure} = 10 \lg \left[ \frac{(S/N)_{\text{in}}}{(S/N)_{\text{out}}} \right] \qquad [10]$$

The reason for specifying the temperature is that most noise is thermal noise (see Chapter 3) and dependent on the temperature. The noise power is proportional to the temperature when specified on the kelvin scale.

For a system with its input at the standard temperature $T_0$ the noise input will be $KT_0$, where $K$ is a constant. The input signal-to-noise ratio is thus $S_{\text{in}}/KT_0$. The noise added by a system can be described by considering the system to be noiseless and to have an extra input to the system at some temperature $T_e$ such that the output from the system is the same as that of the system with its internal noise. If the system has a gain of $G$ then the output signal is $GS_{\text{in}}$ and the output noise $G(KT_0 + KT_e)$. Thus, using equation [9], the noise factor $F$ is

$$F = \frac{S_{\text{in}}/KT_0}{GS_{\text{in}}/G(KT_0 + KT_e)}$$

$$F = \frac{T_0 + T_e}{T_0} \qquad [11]$$

Noise figures are usually measured by injecting noise power from a noise source at two different temperatures into the device for which the measurement is being made and measuring the output at these two temperatures. If the noise power output is $N_0$ when the noise source is at the standard temperature $T_0$ and $N_1$ when it is temperature $T_1$, then

$$N_0 = K(T_0 + T_e)$$

and

$$N_1 = K(T_1 + T_e)$$

The ratio of $N_1$ to $N_0$ is called the *Y-factor* of the system being tested. Thus

$$Y = \frac{N_1}{N_0} = \frac{T_1 + T_e}{T_0 + T_e}$$

Rearranged, this can be written as

$$T_e = \frac{T_1 - YT_0}{Y - 1}$$

Substituting this in equation [11] gives

$$F = \frac{T_0 + [(T_1 - YT_0)/(Y - 1)]}{T_0} = \frac{T_0(Y - 1) + (T_1 - YT_0)}{T_0(Y - 1)}$$

$$= \frac{(T_1/T_0) - 1}{Y - 1}$$

The noise figure is thus

$$\text{noise figure} = 10\lg[(T_1/T_0) - 1] - 10\lg(Y - 1) \qquad [12]$$

The term $10\lg[(T_1/T_0) - 1]$ is called the *excess noise ratio* of the noise source.

Figure 14.13 shows the arrangement that can be used to measure the noise figure. With the noise source cold, i.e., at the standard temperature of 290 K, the noise power is measured using a true r.m.s. meter such as a thermocouple meter. Such a meter reads the r.m.s. value regardless of the waveform. Then with the noise source hot, at some temperature $T_1$ the measurement is repeated. Hence using equation [12] the noise figure can be calculated.

This system can be automated so that a gating signal is used to switch the noise source on and off and the ratio of the output powers from the system under test computed and displayed on a meter. Such an arrangement is called a noise figure meter.

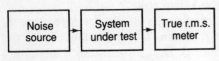

**Fig. 14.13**  Noise measurement

## Problems

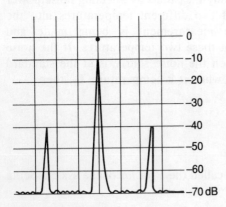

**Fig. 14.14**  Problem 3

1  Describe the principles of a frequency selective voltmeter that can be used for measurements in the audio band of frequencies.
2  Explain the differences in operation and performance between the swept frequency and parallel filter bank forms of spectrum analysers.
3  Figure 14.14 shows the display obtained with an amplitude-modulated sinusoidal signal. The central vertical line of the display has been set to a frequency of 10 MHz and the frequency scan at 2 kHz per horizontal scale division. What are (a) the frequencies of the carrier wave and its sidebands and (b) the modulation index?
4  Figure 14.15 shows the display obtained with an amplitude-modulated sinusoidal signal. The central vertical line of the display has been set to a frequency of 1 MHz, the frequency scan at 5 kHz per horizontal scale division. What are (a) the frequencies of the carrier wave and its sidebands and (b) the modulation index?
5  A spectrum analyser indicates that the output from an amplifier contains the following signals: fundamental $V_1 = 1.0\,\text{V}$ r.m.s., second harmonic $V_2 = 0.015\,\text{V}$ r.m.s., third harmonic $V_3 = 0.008\,\text{V}$ r.m.s., and fourth harmonic $V_4 = 0.003\,\text{V}$ r.m.s. What is (a) the percentage harmonic distortion of the third harmonic and (b) the percentage total harmonic distortion?

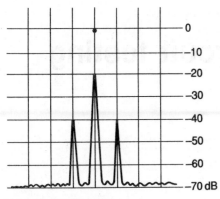

**Fig. 14.15**  Problem 4

6   Explain how the noise figure of an amplifier can be measured.

# 15 Digital circuit testing

## Introduction

The analogue and digital instruments considered in earlier chapters were for the measurement of analogue input signals. This chapter is concerned with instruments for use with digital circuits where the measurement is of the logic state of inputs, outputs or other circuit points. A digital signal has two discrete states: logic 0 represented by a low voltage and logic 1 represented by a high voltage. The values of the voltages depend on the type of logic system used. For TTL (transistor–transistor logic) circuits, logic 0 is 0 V with the threshold for change beginning at 0.8 V and logic 1 is represented by 5 V with the threshold of 2 V. For CMOS (complementary metal oxide semiconductor) circuits, logic 0 is 0 V with the threshold for change beginning at $0.3 \times$ the supply voltage and logic 1 may vary between 3 V and 15 V with the threshold at $0.7 \times$ the supply voltage.

The instruments covered in this chapter range from those, such as probes and pulsers, designed to measure the logic state at a single point in individual integrated circuit devices to comprehensive testers, such as signature analysers and logic analysers, designed for use with microprocessor-based systems.

## Probes, pulsers and tracers

A *logic probe* is an inexpensive hand-held device (Fig. 15.1), shaped like a pen, which can be used to determine the logic level at any point in a circuit to which the probe tip is connected. Indicator lights, light emitting diodes (LEDs), on the body of the probe indicate whether the probe tip is in contact with high or low logic levels. A switch on the probe enables it to be set for different logic levels such as TTL or CMOS. The circuitry for the probe is contained within the body of the probe with a cable from the rear of the probe being used to connect it to the d.c. supply of the circuit under test.

Figure 15.2 shows the basic circuit elements of the probe.

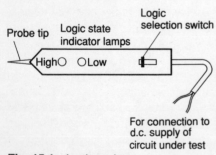

Probe tip | Logic state indicator lamps | Logic selection switch

High○ ○Low

For connection to d.c. supply of circuit under test

**Fig. 15.1** Logic probe

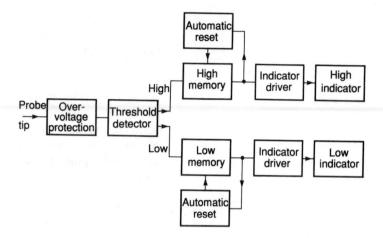

**Fig. 15.2**  Logic probe

The input to the probe is fed through an over-voltage protection circuit to the threshold detector. This is set to the appropriate levels by switching to TTL or CMOS. When the threshold level is detected, the state is fed to the appropriate memory. This retains the state and keeps the indicator on until it is cleared by resetting the memory. The memory and reset circuitry function as a pulse-stretching unit in that pulses with widths as low as 10 ns are able to keep the indicator on for as much as 50 or 100 ms, this being the time between reception of the pulse and resetting. A train of pulses with a frequency less than about 100 Hz will cause both the indicators to blink at about 10 Hz. Wide pulses or square waves above 100 Hz will cause both indicators to glow dimly.

Some probes have only one indicator lamp, high logic being then indicated by the lamp on and low logic by it being off. An open circuit or an input which is between the threshold levels results in a half brightness lamp.

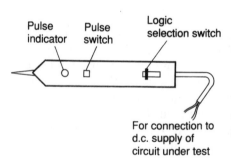

**Fig. 15.3**  Logic pulser

The *logic pulser* (Fig. 15.3) is a hand-held pulse generator, again shaped like a pen, that acts as a signal source for the testing of logic elements. The pulser probe tip is pressed against a node in the circuit and the push button pulse switch on the pulser pressed. The result is an output pulse with a width which is narrow enough, typically less than 300 ns, to limit the amount of energy delivered to the circuit and ensure there is no damage. The outcome is that the node is driven into a high logic state if it was low and a low logic state if it was already high. Used together with the logic probe it is a useful way of checking the functions of logic circuits.

The *current tracer* is similar to the logic probe but it senses pulsing current flow in a circuit rather than voltage levels. The tip senses the magnetic field produced by a current and an indicator light, an LED, turns on when a current flow is detected. The brightness of the LED is used as an indicator of the size of the current. The tracer can be used with a logic pulser to trace current paths, e.g., short-circuits to earth or between lines.

**Logic clips**

A *logic clip* (Fig. 15.4) is a device which clips to an integrated circuit (IC) and makes contact with each of the IC pins. The logic clip has its own logic network to enable it to distinguish between the various pins, drawing its power from the appropriate pins. The logic state of each pin is then shown by LED indicators, there being one for each pin.

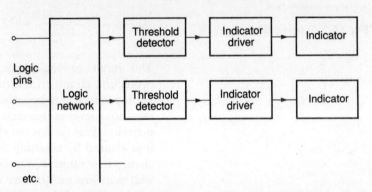

**Fig. 15.4** Logic clip

**Logic comparators**

The logic comparator tests ICs by comparing them to a good IC of the same type. The logic state of a test IC pin is compared with that of the good IC and the test repeated for all the pins. If the logic states are the same then there is no output signal and the LED indicator does not come on. If, however, the states are different then the LED is switched on and so indicates a faulty pin. Figure 15.5 shows the basic form of the comparator; for simplicity, the diagram only shows the comparison for one of the test IC pins. Similar arrangements

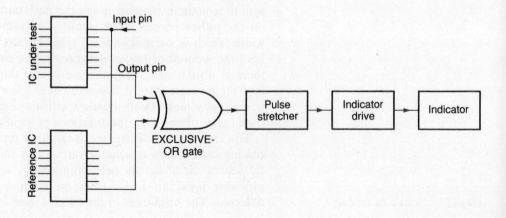

**Fig. 15.5** Logic comparator

exist for each of the pins. Each input pin of the test IC is connected in parallel with the reference IC pin so that both receive the same input signals. The output from a test IC pin is then compared with that from the reference IC pin by an EXCLUSIVE-OR gate which gives an output when the two are different. The pulse stretcher is used to extend the duration of the signal fed to the indicator so that very short duration pulses will result in the indicator being on for a noticeable time.

## Signature analysis

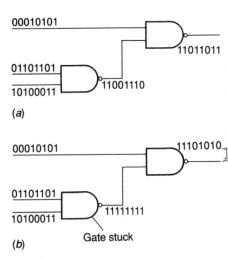

(a)

(b)

Gate stuck

**Fig. 15.6** Node signals: (a) correct, (b) with stuck gate

With analogue systems fault-finding usually involves tracing through the circuitry and examining the waveforms at various nodes. In analogue electronics the waveforms, with their frequency and amplitude, are readily identifiable. Comparison of the waveforms present with what would be expected enables faults to be identified and located. With digital systems the procedure is more complex since trains of pulses at nodes all look very similar to each other. For example, the two NAND gates shown in Fig. 15.6(a) might correctly have the inputs and node signals shown. However, faults may be caused, for example, by a gate sticking at logic 0 or at logic 1. The result might thus be that the node signals become as in Fig. 15.6(b), as a result of one of the gates sticking at logic 1. The problem is thus identifying whether the sequence of pulses at a node is the correct one. To identify whether there is a fault the sequence of pulses at a node has to be converted into a more readily identifiable form. This involves compressing the binary code designation of the pulses into a more compact form such as a small number of hexadecimal digits. This compressed measure of the pulse sequence is called the *signature*. This can then be more readily compared with what should be expected and so enable faults to be identified.

Figure 15.7 shows the basic form of a signature analyser. Start and stop pulses are used to specify the window during which the probe reading is observed. The probe reading is then compared with a clock and the logic state determined at some point on the clock pulses, e.g., the leading edges (Fig. 15.8). The result of this is a data sample. This is then fed into

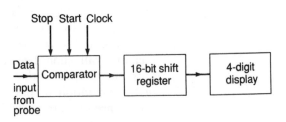

**Fig. 15.7**  Signature analyser

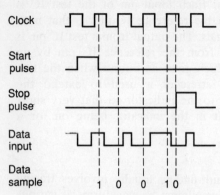

Clock

Start pulse

Stop pulse

Data input

Data sample     1    0    0    1 0

**Fig. 15.8** Data sampling

a 16-bit shift register which converts the data sample into a 16-bit sequence. This conversion puts the data into a form which ideally will yield a short signature for the faulty system that will differ from that given by a fault-free system, even when there is just a single bit error. The output from the 16-bit register is then compressed to give a 4-digit display. The 4-digit displays tend to use hex notation. This is counting to the base 16, i.e., hexadecimal. Table 15.1 shows the data compression codes used with this system. A data sample of, say, 1000110101010001 has four 4-bit sequences, namely 1000, 1101, 0101 and 0001, and these become the display 8D51.

**Table 15.1** Hex

| Bit sequence | Code |
| --- | --- |
| 0 0 0 0 | 0 |
| 0 0 0 1 | 1 |
| 0 0 1 0 | 2 |
| 0 0 1 1 | 3 |
| 0 1 0 0 | 4 |
| 0 1 0 1 | 5 |
| 0 1 1 0 | 6 |
| 0 1 1 1 | 7 |
| 1 0 0 0 | 8 |
| 1 0 0 1 | 9 |
| 1 0 1 0 | A |
| 1 0 1 1 | B |
| 1 1 0 0 | C |
| 1 1 0 1 | D |
| 1 1 1 0 | E |
| 1 1 1 1 | F |

**Logic analysers**

The logic analyser is used to record simultaneously the logic level at many points on a unit under test. Typically, this might mean 8, 16, 32, 48 or more inputs, known as channels. The data are not continuously recorded but sampled, the sampling rate being determined by a clock and being, for example, on every positive rising clock pulse edge. The sampled data are stored in digital form in a memory. With a multi-channel operation and a high sampling rate this could result in a lot of information being required to be stored in the memory. With, for example, 16 channels and a clock running at 20 MHz, 320 million samples will be taken each second. Even a large memory will soon become full. For this reason, selective triggering is used so that data are only stored for that portion of the operation of interest. This may be triggering on the occurrence of a predetermined data word, i.e., when the input signal from a particular input matches the data word inputted from a keyboard by an operator. The term *word* is used for a set of binary numbers. Pre- and post-triggering can be used to store the data immediately before and after the inputted word. Further refinement may be multiple triggering, i.e., more than one trigger word is used and all of them must be received before data is captured. Another possibility is sequential event triggering when a sequence of words is needed before data are captured.

Figure 15.9 shows the basic blocks of a logic analyser. The input signals are connected into the analyser via *pods*. These are interface modules which are designed to provide the necessary mechanical and electrical connections to the item under test, to pick up the input signals without cross talk, i.e., interference signals from other input lines, and generally format the data in such a way that the analyser can understand it. The data inputs to the pods are compared against a threshold level which sets the logic 0 and logic 1 states. The data are then subject to selective triggering in the pattern and sequence recognition memory. The threshold level and the pattern and sequence recognition memory are controlled by an input from a keyboard to the control unit. The control unit coordinates the overall operation of the instrument, stores the required data in memory and passes them on for display, in the required format, on the screen of a cathode ray tube. The clock used for the control unit may be an internal clock or the clock signal from the unit under test. When the internal clock is used, the analyser is said to be operating in the *asynchronous mode*. The frequency of the internal clock, i.e., the frequency with which sampling occurs, must be at least four times more than that of the timing clock in the unit under test if logic states are not to be missed. When the clock signal from the unit under test is used, the analyser is in the *synchronous mode*. In such a mode the acquisition of data occurs synchronously with the timing clock of the unit under test. The advantage of the asynchronous mode is that data may be sampled at a faster rate and thus there is a greater chance of capturing glitches. *Glitches* are signals due not to data but to some form of corruption of the input line. They show as short duration spikes. A clock frequency of 100 MHz is able to capture and display glitches and pulsed data that are about 10 ns long. The synchronous mode has the advantage that logic state samples are taken for each timing cycle of the unit under test.

One of the main uses of logic analysers is to determine signal inputs as functions of time. The screen display can thus be waveforms on a screen showing logic levels against time, such diagrams being called *timing diagrams*. Time intervals

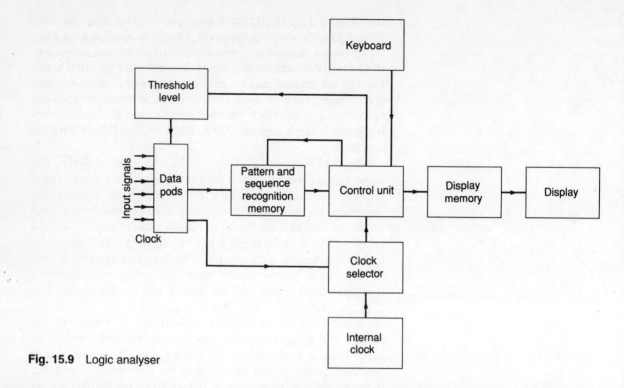

**Fig. 15.9** Logic analyser

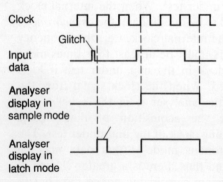

**Fig. 15.10** Glitches

between two points on a timing diagram can be determined by moving a cursor to first one point and then the next. The elapsed time then appears on the screen. Timing diagrams are not always the best form of display when investigating logic states and so a *state list* is often used. This shows the logic states of each input in both binary and hexadecimal form.

With both the asynchronous and synchronous modes of operation data are only stored in the memory at active clock edges, e.g., a positive rising edge. Short duration glitches which occur between clock pulses can thus be missed. To catch such glitches the analyser is operated in what is termed the *latch mode*. In this mode, the sampling frequency has to be greater than the data frequency, i.e., the internal clock is used. Thus if more than one signal edge occurs between two clock edges it can be assumed to be a glitch. Signals between clock edges are thus stored in the memory of a latch circuit and used to trigger off the display of a full clock pulse in the output trace on the screen (Fig. 15.10).

**Problems**

1   Explain how a logic probe can be used to determine the logic state at a point in a digital circuit.
2   A current tracer is used with a pulser to trace an earth fault in a digital circuit. When the tip of the current tracer touches the tip

of the pulser, the tracer LED glows brightly. Now when the tracer tip is moved along a circuit branch A the lamp goes off. When it is moved along branch B it continues to glow brightly. Which is the earth fault branch?

3   Explain the basic method of operation of the logic comparator.

4   Explain what is meant by the signature at a node of a circuit and the principles of signature analysis.

5   Explain how a logic analyser can be used to detect glitches.

6   For a logic analyser explain the differences between (*a*) synchronous and asynchronous operation and (*b*) word, multiple word and sequential event triggering.

# 16 Automatic instruments

Automatic measurement and testing systems are ones where a control unit in the instrument takes over many of the operations that a human operator would have to carry out if using a non-automatic instrument. Thus an automatic voltmeter is, for example, automatically able to select d.c. or a.c. and suitable ranges, indicate the maximum and minimum values of the voltage and carry out statistical analysis, all without the intervention of a human operator.

In general, testing can be considered to involve the application to the unit under test of particular signal waveforms and the observation and evaluation of the resulting output signals (Fig. 16.1). Manual testing involves a human operator carrying out the test procedure, i.e., reading the test schedule and test instructions; selecting, applying and adjusting test input signals; selecting appropriate ranges on, and adjusting, measurement instruments; taking the readings resulting from the measurements; repeating the sequence for each of the tests required; and then evaluating the results. Such a procedure can be automated with automatic testing equipment (ATE), microprocessors exercising the control and data processing roles of the operator. Automatic testing involves connecting the unit under test to the automatic tester via an interface which connects the various points on the unit to the tester. The control unit in the tester then runs the sequence of tests in accordance with the test programme, connecting test signals to the appropriate test points and selecting, and connecting, the appropriate measurement instruments. The tests are carried out, the measurements made and recorded, and then the sequence automatically repeated for each of the tests required. The results of the testing are then automatically compared with the standard pre-programmed values, interpreted and the outcome indicated as perhaps pass or fail.

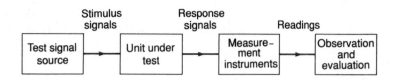

**Fig. 16.1**  The basic test system

**The basic automatic tester**

A test system requires:

1   An interface device to connect the measurement points in the unit under test to the tester and provide suitable signal inputs to the tester.
2   Test signal sources to provide the appropriate input signals to the unit under test.
3   Measurement instruments to measure the responses produced by the unit under test.
4   Switching arrangements to ensure that the appropriate signals and measurement instruments are connected to the appropriate inputs or outputs of the unit under test.
5   A controller to control the signals, instruments, switching, processing of the results, and evaluation of the measurement values by comparison with pre-programmed values.
6   An output from the system to perhaps the screen of a visual display unit (VDU) or in the form of a printout.

Figure 16.2 shows a block diagram of an automatic test system. A commonly used test head fixture to provide the hardware interface between the test system and the unit under test when it is for in-circuit printed circuit board testing is a *bed of nails*. The bed of nails consists of a matrix of spring-loaded pins, each of which makes contact with one of the circuit nodes on the printed circuit board.

ATE may be used at various stages in a product's life: in design and development, in production, in reliability and certification testing, and in service. It may thus be used to test individual components, assembled parts such as printed circuits prior to packaging, packaged assemblies, and in fault

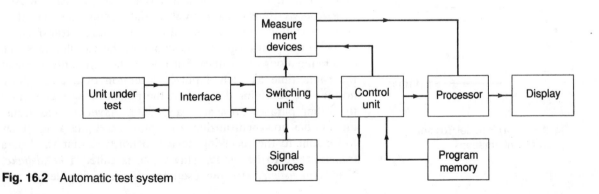

**Fig. 16.2**  Automatic test system

diagnosis and consequent repair of a used product. An example of an ATE system is the logic analyser described in Chapter 15.

## Information transfer

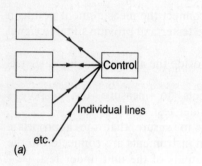

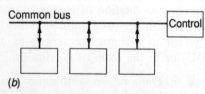

(b)

**Fig. 16.3** Interconnection via (a) individual lines, (b) a bus

Within an automatic measurement or testing system there is a movement of information between the individual constituent units. This information transfer can be to transfer data, send information about the location in the memory where information is to be stored or where it is already stored, or exercise control. Each of the units in the system can be connected via individual cables (Fig. 16.3(a)). This is a comparatively simple approach to connecting units but has the disadvantage that any change or expansion of the system is difficult. An alternative is to use the bus-line system. With this the main hardware elements are connected to a common party line, this being known as a *bus* (Fig. 16.3(b)). A bus is a group of connecting wires or tracks used as a common path for digital information transfer. There are three main buses: the data bus which is used to transfer data between elements, the address bus which is used to send the address of memory locations so that data may be stored or retrieved in addressed locations, and the control bus which is used to send control signals.

Information in automatic measurement and testing systems, and digital computers, is often transmitted as 8-bit words. There are two types of bus system, parallel and serial, which can be used to transfer such information. The *parallel bus* uses eight separate wires to transmit an 8-bit word, each line simultaneously carrying one bit (Fig. 16.4(a)). With the *serial bus* only one line is used to transfer information (Fig. 16.4(b)). The information transfer then takes place bit-by-bit, sequentially, along this line. This method is slower than the parallel method but has the advantage of being cheaper. Parallel buses tend to be used for short distances, up to about 2 m, while serial buses are used for longer distances.

Information can be transmitted along both parallel and serial lines either synchronously or asynchronously. *Synchronous* transmission involves data being transferred in synchronization with a clock pulse. With parallel transmission a clock pulse is transmitted along one of the parallel lines, the data being along other lines. Since valid data will only arrive at the same time as such a pulse, reception of a clock pulse notifies the receiving device that the pulses being received on the other parallel lines are valid data. *Asynchronous* transmission has no synchronizing clock pulses and thus a signal has to be sent to the receiving device to notify it that valid data transfers are to be made. This signal is called a *handshake*. The term *listener* is often used for the receiving device and

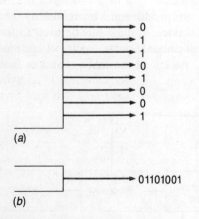

(a)

(b)

**Fig. 16.4** (a) Parallel, (b) serial transfer of information

*talker* for the transmitting device. Reception of a handshake by the listener thus indicates that the talker has the data ready for transmission. When the receiving device, the listener, responds to indicate that it is ready to receive data then the transmitting device, the talker, knows that the receiver is in a state to accept data and so the data word can be transmitted. With parallel transmission, these signals can be sent down individual parallel lines, separate from the data lines.

With a serial bus, data transfers usually take place asynchronously with each data word transferred carrying its own start and stop bits to indicate the start and end of a word. The listener waits for the falling edge of a start bit before accepting serial data. Because not all the bits transferred are data, some being start and stop bits, it is not possible to give a measure of the data transfer rate as the number of bits transferred per second. Instead the data transfer rate is given in terms of the *baud rate*, one bit transferred per second being one baud. Thus

$$\text{baud rate} = \frac{1}{\text{bit time}} \qquad [1]$$

Thus a baud rate of 2400 means that the time taken to transfer one bit is $1/2400 = 417\,\mu s$.

**Standard buses**

A problem that can be encountered in building up an automatic testing system, or indeed any microprocessor-based system, is the multiplicity of different connectors, control signals, logic levels and operating speeds of the constituent elements. Interconnection can then be a nightmare. The problem is, however, simplified if the manufacturers of the elements all adopt a standard bus. Such a standard specifies the type of connector, the signals to be used and their position on the connector. The most commonly used parallel interface bus in automatic testing systems is the IEEE-488 bus (first defined by the American Institute of Electrical and Electronics Engineers, the IEEE). This is also sometimes referred to as the general purpose interface bus (GPIB).

The IEEE-488 bus comprises 16 lines. Eight of these lines are used to carry data such as measurement data, programme data and addresses, three are for handshaking functions and five for bus management functions. Figure 16.5 shows the bus and its use with a bus-connected system involving a controller, listeners and talkers. Up to 15 devices can be interconnected to the bus, provided that the total cable length does not exceed 20 m. Data can be transferred at up to 1 megabyte per second.

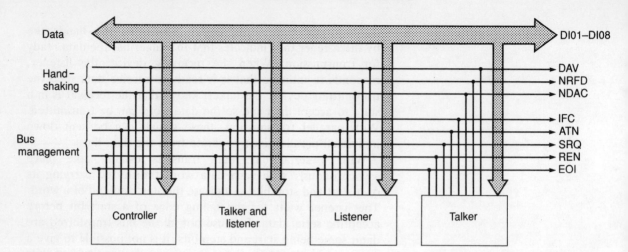

**Fig. 16.5**  IEEE-488 bus connected system

Three lines are for handshaking functions, these lines being 'data valid', 'not ready for data' and 'not data accepted'. The 'data valid' (DAV) line is set low by a talker to indicate that information is ready for reading. The 'not ready for data' (NRFD) line is held low by any listener that is not ready to receive data. The 'not data accepted' (NDAC) line is used by a listener to indicate that the data has been accepted. A typical handshaking sequence between a talker and a number of listeners is: the talker indicates that the data lines carry valid data by making DAV low. A listener that is not ready to receive data would make NRFD low. Data transfer occurs to the other listeners and continues until all the listeners have accepted the data and set NDAC high. The talker responds to this signal by removing the data from the bus and setting DAV high. The listeners respond to this by setting NDAC low to indicate they are ready to receive further data.

Five lines are for bus management functions to ensure an orderly flow of messages. These lines are 'interface clear' (IFC) which resets all devices to a known state, 'attention' (ATN) which alerts all devices on the bus that the data lines contain a command or address, 'service request' (SRQ) which enables any device to signal the controller that it needs attention, 'remote enable' (REN) which selects remote control operation of a device, and 'end or identity' (EOI) which is used either to signify the end of a message sequence from a talker, or after a SRQ signal as a request that the device identifies itself by transferring its own discrete address on to the bus.

**I/O interfacing**

A consequence of using the bus system is that all the devices connected to it must have an input/output interface unit which can convert the data arriving from the bus into a format suitable for operating the device, and/or convert the output from the device into a format suitable for the bus system. Interface units for inputs to the bus system from devices thus might include the functions of matching, protection and signal conditioning. Signal matching might mean converting an input of a resistance change into a d.c. voltage signal, or perhaps an a.c. signal into a d.c. signal. Protection may be series current-limiting resistors or fuses to protect the microprocessor system from high power connections and hence damage. Signal conditioning can involve amplification or attenuation in order to ensure that the signals are the right range, filtering to remove interference, buffering to match impedances and reduce loading errors, offsetting to change the zero of the signal and perhaps remove an unusable part of the signal range, and linearization to obtain a linear relationship between the output from the system being measured and the input to the measurement system. Interface units for outputs from the bus system to devices, such as printers, might include the functions of matching to convert voltage signals to signals needed to operate a device, protection to prevent damaging power levels being connected back into the interface, and signal conditioning.

## A programmable voltmeter

Figure 16.6 shows, as an example of a programmable instrument, the basic form of a programmable digital voltmeter. Such a voltmeter can be programmed by using a keyboard so that it carries out arithmetical operations on inputs, e.g., expresses voltages as a ratio in decibels, carries out some statistical analysis, and lists the number of values above or below or between certain limits. The upper part of the figure represents the input interface with the micro-processor-controlled system. The input unit contains input protection circuits and circuits to select the operations that follow. The amplifier and limit changer gives an output of an appropriate voltage level for the analogue-to-digital converter. The resulting input signal is, for protection purposes, passed to the microprocessor-controlled system via an optical connection. Essentially this is just a light emitting diode (LED) and a photo-sensitive device in close proximity (Fig. 16.7). The input to the connector controls the current flowing through the LED and hence its light output. The photo-sensitive device, a photo-transistor, responds to this light and produces an output signal which replicates the input signal but is physically isolated from it. The digital circuit control is by a bus-connected microprocessor according to instructions received from the keyboard or, though not shown on the figure, from a standard bus interface with some other system. The random

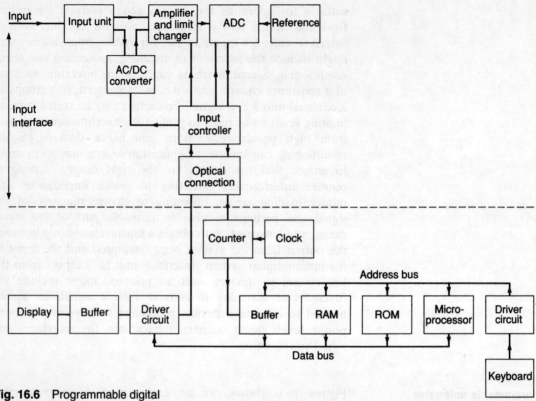

**Fig. 16.6** Programmable digital voltmeter

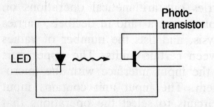

**Fig. 16.7** Optical connector

access memory (RAM) is a temporary memory used to store measurement data during processing. The read only memory (ROM) is a permanent memory which is used to store data necessary for the operation of the instrument.

Programmed instruments have the advantages of offering a more versatile device which can process measurements and give outcomes in a simpler way than is possible using non-programmable instruments. Automatic calibration is feasible, e.g., a calibration check carried out every time the instrument is switched on. Fault diagnosis and repair can, however, be a greater problem than with a non-programmable instrument.

**Problems**

1 Explain the difference between parallel and serial data transmission and state one advantage and disadvantage of each.
2 Explain what is meant by handshaking.
3 Explain the functions of the three lines used for handshaking with the IEE-488 bus.
4 What are the advantages of using a standard bus?
5 With asynchronous serial transmission of data, how does the receiving device know the times at which to sample the data line and obtain valid data?
6 Describe and explain the functions of the constituent elements of automatic testing equipment.

# 17 Instrumentation systems

## Introduction

Instrumentation systems for the acquisition of data tend to consist of a number of components which together are used to make the measurement and display or to record the result (Fig. 17.1). These components are an input device, which receives the input signal and converts it into a form suitable for the next component which is signal conditioning. The term signal conditioning is often loosely used to describe signal conditioning and signal processing. Signal conditioning is when the form of a signal is changed, e.g., from a resistance change to a voltage change, with signal processing being when the signal is made to perhaps the right size, perhaps amplified, without changing its form. Signal conditioning and processing is used to modify the signal so that it can operate the third component which is the display or recorder. Input devices are often referred to as sensors, detectors, pick-ups, probes and, most frequently, transducers. This chapter is a brief overview of transducers, signal conditioners and data transmission.

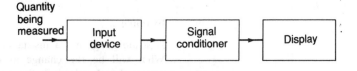

**Fig. 17.1** The general form of measurement systems

## Electrical transducers

The term electrical *transducer* can be used for the input devices for measurement systems which, for the input of the variable being measured, give an electrical output. A stricter definition of a transducer is as components that convert energy or information from one system into energy or information in another system. For example, with a measurement system used to measure temperature, a resistance coil might be used as a resistance thermometer. The resistance of the coil depends on its temperature, a change in temperature producing a change in resistance. Hence information about the temperature of a hot body is transformed into information in the form of resistance. This resistance change can be converted into a current or potential difference change by a Wheatstone bridge, the signal conditioner, and the output displayed as the out-of-balance current. Such a transducer is said to be a *passive device* since it does not itself generate any electrical power. An *active device* is one where the transducer is itself a source of electrical power. An example of this is a thermocouple. As a consequence of the thermoelectric effect, an e.m.f. is generated which is related to the temperature being measured.

In many cases the conversion might be in more than one stage. Thus a load cell used for the measurement of force may consist of a cylinder which is elastically deformed by the action of the force with the deformation then being detected by electrical resistance strain gauges which give a resistance change when subject to strain (see Fig. 9.9). A Wheatstone bridge can then be used to convert the resistance changes into a current or potential difference and so produce a display on a meter.

In order to give some idea of the range of transducer devices and physical principles involved, Table 17.1 and Fig. 17.2 outline some of the more common forms of transducer where the conversion is into an electrical quantity.

### Example 1

What is the change in resistance of a platinum resistance coil of resistance $100\,\Omega$ at $0\,°C$ when the temperature is raised to $30\,°C$? The temperature coefficient of resistance can be taken as $0.0039\,°C^{-1}$.

*Answer*

Assuming that the resistance varies linearly with temperature,

Change in resistance $= R_0\alpha t = 100 \times 0.0039 \times 30 = 11.7\,\Omega$

### Example 2

A strain gauge has a resistance of $120\,\Omega$ and a gauge factor of 2.1. What will be the change in resistance produced if the gauge is mounted with its axis in the direction of a uniaxial strain of 0.0005?

*Answer*

Change in resistance $= GR \times$ strain $= 120 \times 2.1 \times 0.0005$

$= 0.13\,\Omega$

### Example 3

An iron–constantan thermocouple is to be used to measure temperatures between 0 and $400\,°C$. What will be the non-linearity error as a percentage of the full-scale reading at $100\,°C$ if a linear relationship is

**Table 17.1**  Electrical transducers

| Physical principle | Typical applications |
| --- | --- |
| *Resistance* | |
| Resistance of an element depends on the temperature. | Resistance thermometer element such as a resistance wire coil or Thermistor (Fig. 17.2($a$)), for the measurement of temperature. To a reasonable approximation, for a metal wire coil change in resistance = $R_0 \alpha t$, where $R_0$ is the resistance at 0°C, $t$ the temperature in °C and $\alpha$ the coefficient of resistance. For a thermistor $R_t = K e^{\beta/t}$, where $R_t$ is the resistance at $t$°C, $K$ and $\beta$ being constants. Often used with a Wheatstone bridge. |
| Resistance of a wire or semiconductor element depends on the strain. | Strain gauge for the measurement of strain or as a secondary element in some system (Fig. 17.2($b$)). Typically it is a flat coil of wire, which when stuck to a surface gives a resistance change $\Delta R$ proportional to the strain acting on it and hence can be used to determine or give a measure of the strain acting on the surface. $\Delta R/R = G \times$ strain, where $R$ is the gauge resistance and $G$ the gauge factor. A Wheatstone bridge is generally used. |
| The position of the slider of a potentiometer determines that part of its resistance in a circuit. | A p.d. applied across the potentiometer track gives a p.d. between the slider and a track end which is proportional to the slider displacement. Figure 17.2($c$) shows it used as part of a liquid height measurement system. |
| The resistance of photoconductive materials depends on the intensity of light. | Cadmium sulphide is a commonly used material for visible light (Fig. 17.2($d$)). Can be used for the measurement of light intensity or to monitor some other change which is used to chop a beam of light, e.g., the rotation of a sectored disk. |
| *Capacitance* | |
| The capacitance of a parallel plate capacitor depends on the distance between the plates. | This can be used as part of a measurement system for displacement or pressure (Fig. 17.2($e$)). The capacitance $C = \varepsilon_r \varepsilon_0 A/d$, where $\varepsilon_r$ is the relative permittivity, $\varepsilon_0$ the permittivity of free space, $A$ the area of overlap between the two plates and $d$ their separation. |
| The capacitance of a two concentric cylinder capacitor depends on the dielectric between them. | This can be used for measurement of the height of a liquid between the capacitor plates (Fig. 17.2($f$)) with capacitance $C = [2\pi\varepsilon_r\varepsilon_0/\ln(b/a)]$ $[L + (\varepsilon_r - 1)h]$, where $a$ and $b$ are the radii of the two cylinders, $L$ their length and $h$ the height of the liquid above the base of the vertical cylinders. |
| *Inductance* | |
| The reluctance of a magnetic circuit is affected by changes in the magnetic flux path. | Movement of the ferromagnetic plate in Fig. 17.2($g$) affects the flux in the magnetic circuit and hence can affect the e.m.f. induced in a coil. Can be used for measurement of displacement. |
| The inductance of a coil depends on the permeability of its core, hence movement of a high permeability rod into a coil changes its inductance. | Can be used for the measurement of displacement or position. Figure 17.2($h$) shows it as a displacement meter, movement of the plunger determining the relative inductances of the two inductor coils. With the coils in opposite arms of an a.c. bridge the out-of-balance signal is a measure of the position or displacement of the plunger. |
| The difference in the voltages of two secondary coils of a transformer as a result of the position of the iron core. | This is called a linear variable differential transformer and can be used for the measurement of displacement (Fig. 17.2($i$)). |

**Table 17.1** Electrical transducers (continued)

| Physical principle | Typical applications |
|---|---|
| *Thermoelectric*<br>An e.m.f. is produced when there is a difference in temperature between the junctions of two dissimilar metals. | Thermocouple for the measurement of temperature (Fig. 17.2(*j*)). With one junction at 0 °C the e.m.f. $= at + bt^2 + ct^3 + \ldots$, where $t$ is the temperature in °C and $a$, $b$, $c$, $\ldots$, are constants. |
| *Piezoelectric*<br>An e.m.f. is generated when a force is applied to certain crystals, e.g., quartz. | Used in pressure gauges (Fig. 17.2(*k*)), accelerometers, shock and vibration measurement, but only where dynamic measurements are involved. |
| *Photodiodes*<br>A voltage is produced in a semiconductor junction element when radiation is incident on it. | With no illumination a photodiode behaves like a conventional diode, with illumination the diode characteristic shifts (Fig. 17.2(*l*)). The current in reverse bias can be used as a measure of the intensity of illumination. A phototransistor can be considered to be a photodiode connected to an amplifying transistor. |

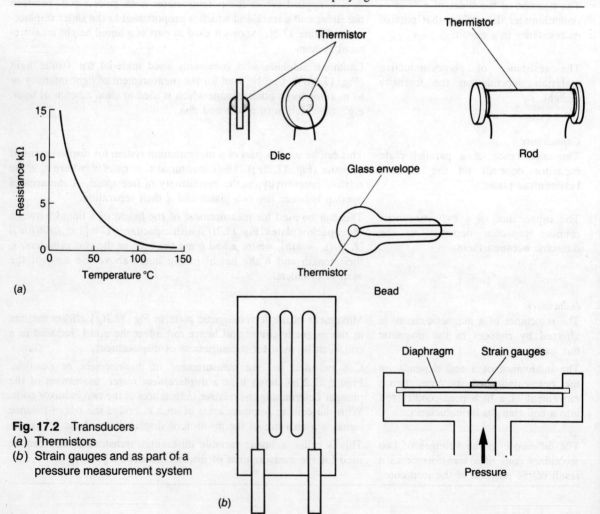

**Fig. 17.2** Transducers
(a) Thermistors
(b) Strain gauges and as part of a pressure measurement system

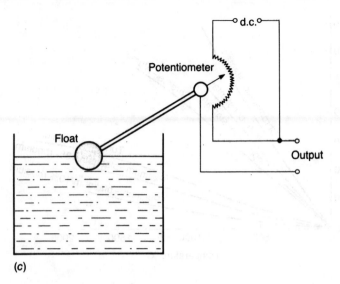

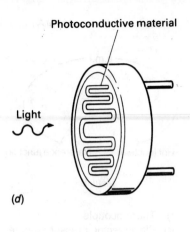

(d)

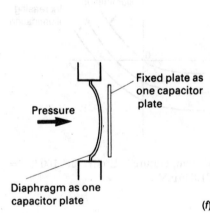

(e)

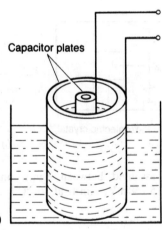

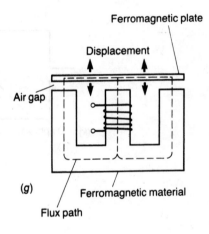

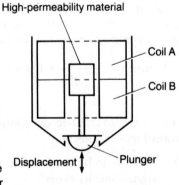

(c) Potentiometer used with liquid
    height measurement system
(d) Photoconductive cell
(e) Capacitative pressure gauge
(f) Capacitative liquid level gauge
(g) Variable reluctance transducer
(h) Variable differential inductor
(i) Linear variable differential
    transformer

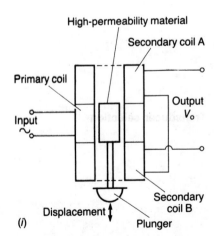

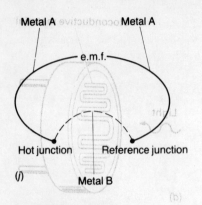

(*j*)

Metal A    Metal A

e.m.f.

Hot junction    Reference junction

Metal B

(*j*)  Thermocouple
(*k*)  Piezoelectric pressure gauge
(*l*)  Photodiode

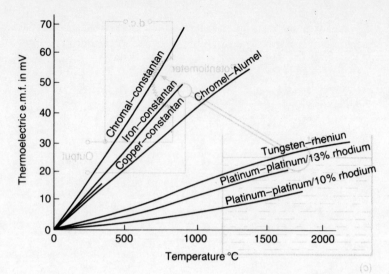

(*k*) Diaphragm    Piezoelectric crystal

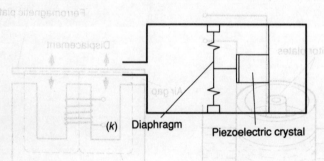

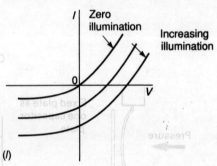

(*l*)

assumed between e.m.f. and temperature? E.m.f. at 100°C = 5.268 mV, e.m.f. at 400°C = 21.846 mV

*Answer*

If there was a linear relationship then the e.m.f at 100°C would be one quarter of that at 400°C, i.e., 5.4615 mV. This is the e.m.f. for which the assumed linear thermocouple indicates 100°C. The actual e.m.f. is 5.268 mV at 100°C and so there is an error of −0.1935 mV. As a percentage of the full-scale reading

$$\text{error} = -\frac{0.1935}{21.846} \times 100 = -0.89\%$$

**Transducer selection**

There are four basic questions that need be posed before a transducer is selected:

1  What is to be the input to the transducer, i.e., the type of signal and its range?
2  What are the required output characteristics from the transducer for it to be compatible with the rest of the measurement system?

**3** What characteristics, i.e., accuracy, sensitivity, resolution, repeatability, linearity, bandwidth, power requirements, cost, etc., are required of the transducer?

**4** What are the environmental conditions, e.g., temperature, moisture, corrosive chemicals, the transducer will have to withstand?

Suppose a transducer is required which has an input of temperature and an output of some electrical quantity. Possibilities include metal resistance coils, thermistors and thermocouples. The following are some of the key characteristics. The temperature range and sensitivity of metal resistance coils depend on the metal used, e.g., nickel has a range of $-80$ to $320\,°C$, copper $-200$ to $260\,°C$, and platinum $-200$ to $850\,°C$. Sensitivity is typically between about 0.1 and $10\,\Omega/°C$. Non-linearity is generally less than about 1% of the full-scale range, stability is a drift of no more than 0.01% per year and accuracy is typically about $\pm0.75\%$. Metal resistance coils are invariably bulky and so no use for measuring highly localized temperatures or, because of their mass and consequently large thermal capacity, rapidly changing temperatures. Thermistors give much larger resistance changes per degree, typically $0.1\text{--}1.0\,k\Omega/°C$, than metal wire coils and have temperature ranges generally within about $-100$ to $350\,°C$. They have a high non-linearity. Over a small range the accuracy can be as good as $0.1\,°C$; however, the stability is a drift of about 1% per year. Unlike the metal coil and thermistor, which give resistance changes, the thermocouple gives an e.m.f. The range and sensitivity depend on the metals used, e.g., chrome–constantan 0 to $980\,°C$ and $63\,\mu V/°C$, iron–constantan $-180$ to $760\,°C$ and $53\,\mu V/°C$, platinum–platinum/rhodium 13% 0 to $1759\,°C$ and $8\,\mu V/°C$. Base metal thermocouples are relatively cheap but deteriorate with age. They have accuracies of the order of $\pm1\text{--}3\%$. Noble metal thermocouples are more expensive but more stable with accuracies of the order of $\pm1\%$ or less. The very small area of a thermocouple junction and its small thermal capacity means that thermocouples can be used for the measurement of highly localized and rapidly changing temperatures.

A common form of pressure transducer is a diaphragm which is supported round the edges with the centre bowing in or out according to the pressure difference between the two sides of the diaphragm. This movement of the diaphragm can be monitored by means of strain gauges (as in Fig. 17.2($b$)), by capacitance changes (as in Fig. 17.2($e$)), by reluctance changes, or by means of a piezoelectric crystal (as in Fig. 17.2($k$)). Strain gauge transducers may employ metal wire or foil strain gauges stuck to the surface of the diaphragm or use a diaphragm made of silicon with semiconductor strain gauges produced in the diaphragm by suitable doping of it. Typical ranges and accuracy of the above are: metal wire strain gauges, range $10^4$–$10^8$ Pa, accuracy $\pm 0.1\%$; integral semiconductor strain gauges, $10^4$–$10^6$ Pa, accuracy $\pm 0.5\%$; capacitative, $10^3$–$10^5$ Pa, accuracy $\pm 0.1\%$; reluctance 1–$10^8$ Pa, accuracy $\pm 0.1\%$; piezoelectric, 0.1–$10^5$ Pa, accuracy $\pm 0.2\%$. Strain gauge, capacitative and reluctance transducers can respond to pressure frequencies from 0 to about 1 kHz. The piezoelectric transducer can only be used for dynamic pressures and has the frequency range 1–100 kHz.

## Signal conditioning and processing

The term *signal conditioning* is used for the element of a measurement system which converts the signal from the transducer into a form suitable for further processing and hence display. The output from a signal conditioner is usually a d.c. voltage or a d.c. current. Typical signal conditioners are bridges where a change in resistance, capacitance or inductance is converted into an out-of-balance potential difference or current. The term *signal processing* is often used for the processes which the signal then undergoes in order to make it suitable to operate the display. This could be amplification, filtering, linearization, offsetting or buffering. Linearization is the process of making the output to the display become proportional to the input to the transducer. Offsetting is the shifting of the zero of the signal so that only that part of the signal of interest results in a display. Buffering is the matching of the impedances so that there is a maximum transfer of energy.

## Operational amplifier

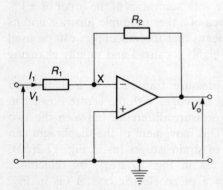

**Fig. 17.3** Inverting amplifier

The operational amplifier is the basic building block for both d.c. and a.c. instrument amplifiers. It has two inputs, known as the inverting input $(-)$ and the non-inverting input $(+)$, and an output. The output to input ratio, i.e. gain, depends on the connections made to these inputs. Figure 17.3 shows the connections made to the operational amplifier when it is used as an *inverting amplifier*. The input is connected to the inverting input through a resistor $R_1$ and the non-inverting input is connected to ground. A feedback path is provided from the output terminal, via a resistor $R_2$, to the inverting terminal.

The operational amplifier itself has a very large gain, 100 000 or more, and the change in its output voltage is generally limited to about $\pm 10$ V. With this gain the input voltage must be between + or − 0.0001 V. This is virtually zero and so point X is at virtually earth potential. For this reason it is called a virtual earth. The potential difference across $R_1$ is $(V_I - V_x)$, hence the input potential $V_I$ can be considered to be across $R_1$, and so

$$V_I = I_1 R_1$$

The operational amplifier has a very high impedance and so virtually no current flows through X into it. Hence the current through $I_1$ flows on through $R_2$. Because X is the virtual earth then, since the potential difference across $R_2$ is $V_x - V_o$, the potential difference across $R_2$ will be virtually $-V_o$. Hence

$$-V_o = I_1 R_2$$

$$\frac{V_o}{V_I} = -\frac{R_2}{R_1} \tag{1}$$

The ratio of the output to input is thus determined by the relative values of $R_2$ and $R_1$. The negative sign indicates that the output is inverted, i.e., 180° out of phase, with respect to the input.

Figure 17.4 shows the operational amplifier connected as a *non-inverting amplifier*. With this circuit the output voltage is in phase with the input voltage and

$$\frac{V_o}{V_I} = \frac{R_2 + R_3}{R_3} \tag{2}$$

When the feedback resistor $R_2 = 0$ then the amplifier has a gain of 1. In this ideal case, the input impedance is infinite. Such a circuit thus allows a very high input impedance to be realized.

Figure 17.5 shows the operational amplifier connected as a *differential amplifier*. Then

$$V_o = \frac{R_2}{R_1}(V_2 - V_1) \tag{3}$$

Such an amplifier finds a use in bridge circuits by amplifying the out-of-balance difference in potential.

With many transducers and signal conditioners the low level output signal may be superimposed on a large common mode signal. This is, for example, the situation with a Wheatstone bridge where the output is a small difference between two larger potentials. Using the operational amplifier circuit given in Fig. 17.5 would thus lead to amplification of both the common mode signal and the superimposed transducer signal. For this reason an *instrumentation amplifier* is of the form shown in Fig. 17.6 with three operational amplifiers being used. Amplifiers 1 and 2 act as a high impedance input stage to amplifier 3 which is a differential amplifier. The differential gain is

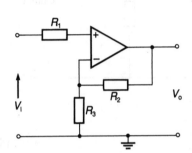

**Fig. 17.4** Non-inverting amplifier

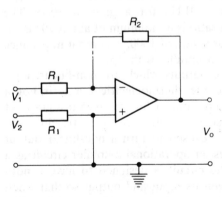

**Fig. 17.5** Differential amplifier

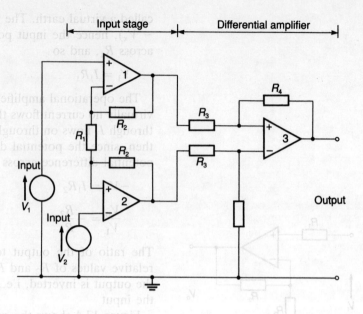

**Fig. 17.6** Instrumentation amplifier

$$\text{gain} = \frac{R_1 + 2R_2}{R_1} \qquad [4]$$

Typically such an amplifier will have a higher input impedance, and a common mode rejection ratio about ten times larger, than a single amplifier. Gain is generally variable between 1 and 1000 with a bandwidth of more than 2 MHz for a gain of 1 which reduces to about 40 kHz for a gain of 1000. The common mode rejection ratio is a minimum of about 70 dB for a gain of 1 and 100 dB for a gain of 1000. The input impedance is of the order of $10^9 \, \Omega$ in parallel with 2 pF.

Some transducers have outputs which are non-linear, e.g., the thermocouple where the thermoelectric e.m.f. is not a linear function of the temperature (this means that a graph of the e.m.f. plotted against temperature is not a straight line). One way that can often be used to turn a non-linear output into a linear one involves an operational amplifier circuit as a *linearization element*. The circuit is designed to have a non-linear relationship between its input and output so that when its input is non-linear its output is linear. This is achieved by suitable choice of components for the feedback loop.

Figure 17.7 shows such a circuit with a diode in the feedback loop. The diode has a non-linear characteristic and this results in the relationship between the input voltage $V_I$ and the output voltage $V_o$ becoming

$$V_2 = C \ln(V_I)$$

with $C$ being a constant. The amplifier is thus non-linear.

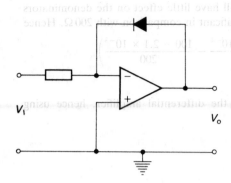

**Fig. 17.7**  Signal linearization circuit

However, if the input $V_I$ is provided by a transducer for its input $\theta$ where

$$V_I = K e^{\alpha\theta}$$

with $K$ and $\alpha$ being constants. Then

$$V_2 = C\ln(Ke^{\alpha\theta})$$
$$= C\ln(K) + C\alpha\theta \qquad [5]$$

The result is a linear relationship between the output $V_2$ and the input to the transducer $\theta$.

**Example 4**

An inverting amplifier has a resistance of $1\,\text{M}\Omega$ in the inverting input line and a feedback resistance of $10\,\text{M}\Omega$. What is the amplifier gain?

*Answer*

Using equation [1]

$$\text{gain} = -\frac{R_2}{R_1} = -10$$

**Example 5**

For the four active arm strain gauge bridge shown in Fig. 9.9, the gauges have a gauge factor of 2.1 and a resistance of $100\,\Omega$. When used with the load cell in the way shown in the figure, an applied load produces a strain of $1.0 \times 10^{-5}$ in the compression gauges and $0.3 \times 10^{-5}$ in the tensile gauges. The supply voltage for the bridge is $10\,\text{V}$. What will be the ratio of the feedback resistor resistance to that of the resistors in the two inputs of a differential amplifier used to amplify the out-of-balance potential difference if the load is to produce an output of $1\,\text{mV}$?

*Answer*

The change in resistance of a gauge subject to the compression is a decrease in resistance and is given by (see Table 17.1)

$$\text{change in resistance} = RG \times \text{strain}$$
$$= -100 \times 2.1 \times 1.0 \times 10^{-5}$$
$$= -2.1 \times 10^{-3}\,\Omega$$

The change in resistance of a gauge subject to the tension is an increase in resistance of

$$\text{change in resistance} = +100 \times 2.1 \times 0.3 \times 10^{-5}$$
$$= +6.3 \times 10^{-4}\,\Omega$$

The out-of-balance potential difference is given by (see Chapter 9, equation [4])

$$V_{Th} = V_s\left(\frac{R_1}{R_1 + R_2} - \frac{R_3}{R_3 + R_4}\right)$$

The changes in resistance will have little effect on the denominators of the equation, being insignificant in comparison with $200\,\Omega$. Hence

$$V_{\text{Th}} = 10\left(\frac{100 + 6.3 \times 10^{-4}}{200} - \frac{100 - 2.1 \times 10^{-3}}{200}\right)$$

$$= 1.4 \times 10^{-4}\,\text{V}$$

This becomes the input to the differential amplifier, hence using equation [3]

$$V_o = \frac{R_2}{R_1}(V_2 - V_1)$$

with $V_o$ to be $1\,\text{mV}$ then

$$1.0 \times 10^{-3} = \frac{R_2}{R_1} \times 1.4 \times 10^{-4}$$

and so $R_2/R_1$ has to be 7.1.

## Analogue-to-digital conversion

Signal processing may involve the conversion to digital form of an analogue input signal for a digital display or so that further processing can occur before display. Such a conversion is likely to involve two elements, a sample-and-hold element and an analogue-to-digital converter (ADC) (Fig. 17.8). The *sample-and-hold* element takes a sample of the analogue signal and holds it long enough for the ADC to convert it to a digital signal. Essentially it is a capacitor which, when switched in parallel with the input during the sampling time, is charged to the analogue voltage. This then holds the sample for the duration of the conversion to digital form (Fig. 17.9).

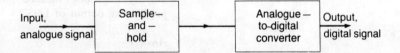

**Fig. 17.8** Analogue-to-digital conversion

The following are some terms commonly used in the specifications of sample-and-hold devices. The *acquisition time* is the time taken for the device to reach the value of the input signal, i.e., the capacitor to become fully charged. A low cost device will have an acquisition time of the order of $10\,\mu\text{s}$, a high speed one $25\,\text{ns}$ or less. The *aperture time* is the time delay that occurs between the time at which the command is given to hold the sample and the capacitor becoming isolated from the analogue input. The signal in the hold mode will *droop*, this term being used to describe the decrease in value per unit time. The output from the sample-and-hold element is converted into a digital signal by the ADC. A low cost 8-bit ADC might have a conversion time of about $20\,\text{ms}$, a fast one about $5\,\mu\text{s}$. Thus the total time taken, with a low cost sample-

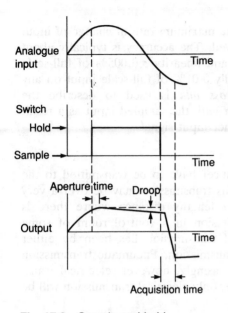

Analogue input — Time

Switch
Hold →
Sample → — Time

Aperture time — Droop

Output — Time

Acquisition time

**Fig. 17.9** Sample-and-hold

## Data acquisition systems

and-hold element plus an 8-bit ADC, to sample the signal and produce a digital output is of the order of $10\,\mu s + 20\,ms$. For rapidly changing analogue signals the total conversion time is an important parameter in determining the sample rate and the extent to which the digital signal will keep track with it. If the ADC has an $n$-bit output then for it to be able to resolve the input analogue signal to within one least significant bit during the total time $T$ it takes for the conversion, the maximum rate of change of the analogue signal with time, the so-called *slew rate*, must be

$$\text{max. rate} = \frac{2^{-n}V}{T} \qquad [6]$$

where $V$ is the full-scale output voltage. For slow sampling the important parameter is the droop rate.

*Data loggers* are data acquisition systems that can take inputs from a variety of sources, carry out some mathematical operations on the inputs and then provide storage for the data in a semiconductor memory or a magnetic tape/disk system. A data logger consists essentially of a multiplexer, a sample-and-hold element, an analogue-to-digital converter and then some means of recording or manipulating and recording the output (Fig. 17.10). The input signals, after perhaps some suitable signal conditioning, are fed into the multiplexer. The multiplexer selects one signal. The sample-and-hold element then takes a sample of the signal and holds it long enough for the analogue-to-digital conversion to take place without error due to input fluctuations. The output from the unit is thus a digital signal. The multiplexer can be switched to each input signal in turn and thus digital outputs obtained for each. These outputs are often fed to a computer which can not only store the data but can also process it.

Typically a data logger may handle 20–100 inputs, though some used with a computer may handle many more. It might have a sample and conversion time of $10\,\mu s$ and a slew rate of

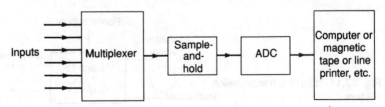

Inputs → Multiplexer → Sample-and-hold → ADC → Computer or magnetic tape or line printer, etc.

**Fig. 17.10** Data logger

0.5 V/s. The *slew rate* is the maximum rate of change of input voltage that can be followed. The accuracy is typically about 0.01% of full-scale input and linearity ±0.005% of full-scale input. Cross talk is typically 0.01% of full-scale input on any one input. The term *cross talk* is used to describe the interference that can occur with the sampled input as a result of the existence of the other input signals.

**Data transmission**

The signals from a transducer have to be transmitted to the point of measurement. This transmission may be over a very short distance or quite a lengthy distance where there is perhaps centralized supervision in a control room of some plant. Traditionally signal transmission has been by either pneumatic or electrical transmission. Pneumatic transmission is by pressurized air. Increasingly, however, electrical transmission is being used; here, only electrical transmission will be considered.

The transducer output signals are usually analogue and they may be retained in this form for the transmission or converted to digital form before being transmitted. *Analogue transmission* can entail the transducer signal being transmitted directly to the point of measurement without any prior signal conditioning. This technique can only be used when the transmission distances are less than about a metre and the signal levels are not too small, i.e., more than about 0.1 V. In such situations, the degradation of the signal due to such factors as cable resistance and external interference is generally small enough not to affect the measurement significantly. Where the transmission distance is more than this or the signals smaller, signal degradation can become too large. For this reason, such transmissions are generally in the form of d.c. current signals in the range 4–20 mA with a zero transducer output being converted to the 4 mA current and the maximum output to 20 mA (Fig. 17.11). This is a fail-safe form of transmission since a break in the transmission loop would show as zero current. Such d.c. currents can be transmitted over distances up to about 3 km without excess degradation.

*Digital signals* can be transmitted in either parallel or serial form (see Chapter 16). With parallel form *n* parallel data lines are used to transmit an *n*-bit word, each line carrying

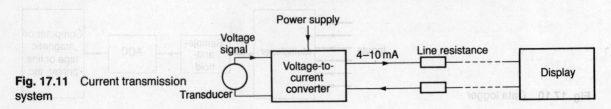

**Fig. 17.11** Current transmission system

simultaneously one bit. With serial transmission only one line is used, the information being transmitted sequentially bit-by-bit. Because of the cost, parallel transmission tends to be used for distances up to about 2 m, with serial transmission being used for longer distances.

Parallel transmission is via a set of parallel lines, referred to as a bus (see Chapter 16). A standard form of bus that is often used for parallel transmission with instruments is the IEEE-488 bus. This comprises 16 lines, 8 being used to transmit measurement data, programme data and addresses, 3 for handshaking functions, i.e., indicating when a message is to be sent and whether the receiver is ready to receive it, and 5 for bus management to ensure an orderly flow of data (see Fig. 16.5).

Serial transmission can be used over very large distances, the pathway varying from a twisted-pair cable to low loss coaxial cable and fibre optic cables. Data from a number of transducers might be selected by a multiplexer for sequential switching for transmission along a line. There are a number of standard methods of serial transmission, the type used depending on the transmission distance, the rate at which data are to be transmitted, and the number of multiplexed signals to be sent over a single line. A common standard used is RS-232C. This specifies a 25-pin plug and socket connection at each end of the transmission link and the signals to be conveyed through each constituent line. Lines are designated for electrical ground, data interchange, control and clock or timing signals. Three lines are used for data transmission, one each for data in each direction and one for a signal ground wire for a common return path. The standard also defines the signal levels, condition and polarity. Data signals are between −3 and −15 V for logic 1 and between +3 and +15 V for logic 0. The maximum transmission rate is 20 000 bit/s and the maximum length of transmission path is limited by the need for the receiver to have no more than 2500 pF across it. Since cables can have capacitances of about 150 pF per metre, this means a maximum transmission distance of about 17 m. A newer RS-449 standard is gradually replacing the RS-232C standard and is capable of higher rates of transmission over longer distances. This standard uses a 37-pin interface with each transmitting and receiving data line having its own separate return line, each data line and return line forming a twisted pair to give shielding from inductively coupled interference.

In order to transmit digital data over long distances modems can be used with the public telephone lines. A *modem* (MOdulator/DEModulator) modulates the digital signal on to an analogue wave form. The modem changes the logic 1 and

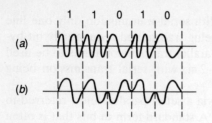

**Fig. 17.12** (a) FSK, (b) PSK

logic 0 into generally either different frequencies, the so-called frequency-shift keying (FSK), or different phases, phase-shift keying (PSK) (Fig. 17.12). Thus the arrangement for sending the digital data is as shown in Fig. 17.13. The data are transmitted in digital form by serial transmission using a RS-232C interface over a short distance to a modem where they are converted into an analogue signal. This is then transmitted down the telephone line to be received by another modem. This converts the analogue signal into a digital signal which passes through a RS-232C interface to the data sink, i.e., the recipient of the data.

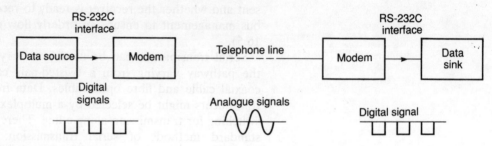

**Fig. 17.13** Data transmission using modems

**Problems**

1 Select and justify your selection of transducers for the following situations:
   (a) Input: rapidly changing temperature.
   (b) Input: light signal; output: resistance change.
   (c) Input: deformation of a diaphragm; output: resistance change.
   (d) Input: rotation of a lever; output: potential difference.

2 Explain the significance of the following extracts from specifications of transducers:

(a) A strain gauge pressure transducer
Range: 2 MPa to 70 MPa, 70 kPa to 1 MPa
Excitation: 10 V d.c. or a.c. (r.m.s.)
Full range output: 40 mV
Non-linearity and hysteresis: ±0.5%
Temperature range: −54 °C to +120 °C
Thermal shift: zero 0.030% full range output/°C
Thermal shift: sensitivity 0.030% full range output/°C

(b) A load cell
Ranges available: 0–50 kg, 0–100 kg, 0–200 kg, 0–500 kg, 0–1000 kg, 0–2000 kg, 0–5000 kg
Total error due to non-linearity, hysteresis and non-repeatability: ±0.25%
Application: low–medium accuracy container weighing

(c) A thermistor
Resistance: 2000 Ω at 20 °C, 40 Ω at 200 °C

Temperature range: $-100$ to $+250\,°C$
Self-heating effect: $1\,°C$ rise in temperature per $1.4\,mW$ power dissipation

(*d*) Electrical resistance strain gauge
Gauge resistance: $100\,\Omega \pm 0.5\%$
Gauge factor: $2.1 \pm 1\%$
Temperature range: $-20\,°C$ to $+60\,°C$
Temperature coefficient of resistance: $<0.05\,°C^{-1}$
Coefficient of expansion: $1.1 \times 10^{-7}\,°C^{-1}$

3 A platinum resistance thermometer has a resistance at $0\,°C$ of $120\,\Omega$ and forms one arm of a Wheatstone bridge. At this temperature the bridge is balanced with each of the other arms also being $120\,\Omega$. The temperature coefficient of resistance of the platinum is $0.0039\,°C^{-1}$. What will be the output voltage for change in temperature of $20\,°C$ if the instrument used to measure it can be assumed to have infinite resistance and the supply voltage, with negligible internal resistance, for the bridge is $6.0\,V$?

4 For the four active arm strain gauge bridge shown in Fig. 9.9, the gauges have a gauge factor of 2.1 and a resistance of $120\,\Omega$. When attached to the diaphragm of a pressure gauge an applied pressure difference from one side of the diaphragm to the other produces a strain of $+1.0 \times 10^{-5}$ in two of the gauges and $-1.0 \times 10^{-5}$ in the other two. The supply voltage for the bridge is $10\,V$. What will be the out-of-balance potential difference output from the bridge assuming the load across the output has effectively infinite resistance?

5 A De Souty a.c. bridge is to be used with the capacitative liquid level gauge described in Fig. 17.2(*f*). The ratio $b/a$ of the diameters of the concentric cylinders is 2.0 and the length of the cylinders $3.0\,m$. If the liquid has a relative permittivity of 2.1, what will be the value of the capacitor required in the bridge to give balance when the liquid level is $1.0\,m$? The resistance in the arm of the bridge opposite the gauge is $100\,\Omega$ and that opposite the capacitor liquid level gauge $10\,k\Omega$.

6 A linear variable differential transformer (LVDT) produces an output of $2\,V$ r.m.s. for a displacement input of $1.0\,mm$. If the voltage is read on a voltmeter, range $0–5\,V$ with a total of 100 scale divisions, what is the displacement corresponding to a change of 1 division on the voltmeter?

7 A cylindrical load cell, of the form shown in Fig. 9.9, has four strain gauges attached to its surface. Two of the gauges are in the circumferential direction and two in the longitudinal axis direction. When the cylinder is subject to a compressive load, the axial gauges will be in compression while the circumferential ones will be in tension. If the material of the cylinder has a cross-sectional area $A$ and an elastic modulus $E$, then a force $F$ acting on the cylinder will give a strain acting on the axial gauges of $-F/AE$ and on the circumferential gauges of $+\nu F/AE$, where $\nu$ is Poisson's ratio for the material.

Design a complete measurement system, using load cells, that could be used to monitor the mass of water in a tank. The tank itself has a mass of 20 kg and the water, when at the required level, 40 kg. The mass is to be monitored to an accuracy of ±0.5 kg. The strain gauges have a gauge factor of 2.1 and are all the same resistance of 120.0 Ω. For all other items, specify what your design requires.

8   A suggested design for the measurement of liquid level involves a float which in its vertical motion bends a cantilever. The degree of bending of the cantilever is then taken as a measure of the liquid level. Strain gauges are used to monitor the bending of the cantilever with two gauges being attached to its upper surface and two to the lower surface. The gauges are then to be incorporated in a four gauge active Wheatstone bridge with the out-of-balance potential difference monitored.

Carry out preliminary design calculations to determine whether the idea is feasible and what values and form the various components may need to take.

When a force $F$ is applied to the free end of a cantilever of length $L$, the strain on its surface a distance $x$ from the clamped end is

$$\text{strain} = \frac{6(L - x)F}{wt^2E}$$

where $w$ is the width of the cantilever, $t$ its thickness and $E$ the tensile modulus of the material. The upthrust force exerted by a float equals the weight of fluid displaced by the float.

9   Explain the significance of the following information taken from the specification of an instrument amplifier.

Gain: 1 to 1000
Bandwidth: gain = 1, >2 MHz; gain = 1000, 40 kHz
Common mode rejection ratio: gain = 1, 70 dB min.
                             gain = 1000, 100 dB min.
Input impedance: differential, $3 \times 10^9 \, \Omega$ in parallel with 2.0 pF
                 common mode, $6 \times 10^{10} \, \Omega$ in parallel with 3.0 pF

10  Explain the significance of the slew rate for a data logger.
11  Explain the factors that have to be considered in determining how data should be transmitted from a transducer to a display system.
12  As part of the central heating control system for a building it is proposed to monitor the temperature at 20 different, widespread, points in the building. The temperatures are envisaged as being within the range 0–30 °C and an accuracy of ±1 °C is required. Propose a system which would enable all the temperatures to be monitored in a central control room.

# 18 Test procedures

**Introduction**

Testing is the taking of measurements to ascertain whether some product conforms to specified standards and quality. Testing can occur during the development of a product to confirm design decisions, during manufacture to control quality, at acceptance by the customer to ensure the product is up to specification, and during its operational life to ensure that it is continuing to perform to specifications and to diagnose faults. The earlier chapters in this book have considered the instruments of testing, here the economics of testing and test procedures are considered.

**The economics of testing**

Testing costs money. However, the absence of testing also costs money in that the failure to detect faulty units during development or manufacture will result in faulty products that sooner or later will have to be put right or rejected. What has to be balanced is the quantity and quality of the testing such that its cost is less than the cost of not testing. The costs of testing can be considered to be:

1 The capital cost of the test equipment.
2 The capital cost of producing the test programmes.
3 The maintenance cost of the test equipment.
4 The operating costs of the test equipment and programmes.
5 Depreciation costs of the equipment.

Against this have to be balanced:

1 The consequential costs of missing a faulty unit, early in the design/manufacturing cycle or late in the cycle.
2 The cost of recovering a faulty unit by virtue of early testing.

There are two main approaches to testing, using manual techniques or using automatic testing equipment (ATE). ATE has higher capital and maintenance costs but is capable of a greater throughput in units tested per hour. This greater throughput can make ATE more cost effective than manual testing where there is high volume production.

**Manual testing**

Manual testing involves a person using test instruments to make measurements on a unit and determine whether it is operating to specification and, if faulty, where the fault is. A test specification to check that a unit gives the required performance might involve:

1 With no power on, carry out continuity and insulation tests. These might be: checking that the negative output terminal is connected to the chassis, checking that the chassis is connected to the mains earth connector, checking that the insulation resistance between the line connector and the chassis earth is greater than, say, $50\,\text{M}\Omega$.

2 With the unit connected to the mains power, carry out voltage and signal measurements. These might be: checking that the output from the unit has the correct values and signal forms for the various settings of the unit (a so-called performance test), measuring the voltage at certain key points in the circuit and checking that it has the required values, checking the signal form at certain key points in the circuit. These measurements might involve the use of a multi-range meter and an oscilloscope.

3 With the unit connected to the mains power, carry out a soak test. This involves checking the performance of the unit when left on for a period of time under specified conditions.

**Manual test plans**

Manual test plans tend to be determined as follows:

1 A list is made of all the important functions and characteristics of the constituent elements of the unit under test and the maximum and minimum values allowed.

2 The test instruments needed to carry out the above measurements are specified.

3 Details of the test points and circuits are specified.

4 The required tests are listed, starting off with those to be made prior to power being applied to the unit, then with those with power connected and finally with signal inputs to the unit.

On the basis of the above, a test sheet can be drawn up, listing the test instruments, the tests to be carried out and their sequence, and the test results which would indicate the unit is correctly functioning.

Thus, for example, in the testing of a power supply the quantities that might be tested are the d.c. output voltage, the d.c. output current available, the output ripple voltage at full load, the stabilization against mains supply changes and the regulation from zero to full load. Figure 18.1 shows the test circuit that might be used for the above measurements. The peak-to-peak ripple amplitude is measured with the oscilloscope. For the stabilization measurement, the unit is fully loaded and the change in d.c. output voltage determined for, say, a ±10% change in the a.c. input. If such a change produced a change in output voltage of, say, 0.5% then the stabilization is quoted as 20/0.5 = 40 to 1. The load regulation involves the measurement of the change in d.c. output when the load is changed from zero to full load, being

$$\text{regulation} = \frac{\text{change in d.c. output}}{\text{d.c. output on no load}} \times 100\%$$

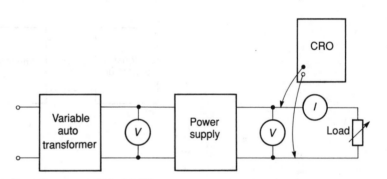

**Fig. 18.1**   Testing a power supply

Testing an amplifier could involve measurements being made of the gain, frequency response and bandwidth, input impedance, output impedance, output power and distortion. The measurement of the amplifier gain is usually carried out by means of a calibrated attenuator. Figure 18.2(*a*) shows the circuit. With the signal generator set to give the frequency at which the gain is required and the attenuator set to zero, the oscilloscope is used to monitor the amplifier output across the load, the trace height being noted. Then the oscilloscope leads are moved to the amplifier input and the attenuator adjusted to give the same height trace on the oscilloscope. The gain of the amplifier is equal to the setting of the attenuator. The same arrangement can be used to determine how the gain varies with the input frequency and hence the bandwidth. The input impedance at low frequencies is largely resistive and can be measured by means of the circuit shown in Fig. 18.2(*b*). With the variable resistor set to zero the trace height on the

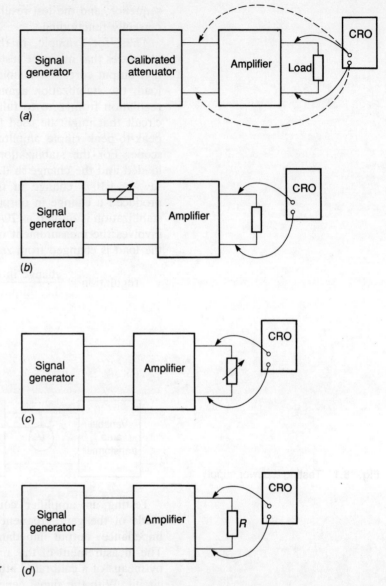

**Fig. 18.2** Testing an amplifier:
(a) gain, (b) input resistance,
(c) output resistance, (d) output power

oscilloscope is noted. Then the variable resistance is increased until the trace height is reduced by half. This means that the oscilloscope is across just half the potential difference, the other half being across the resistor. Thus the input resistance is the value of the variable resistor. The output impedance can be measured in a similar way (Fig. 18.2(c)). The variable resistance in the output circuit is set to zero and the trace height measured; then the resistor is increased until the trace is reduced by half. The output resistance is the value of the variable resistance. The power output of the amplifier can be measured by the circuit shown in Fig. 18.2(d). The oscilloscope is used to determine the r.m.s. value of the output voltage $V_{rms}$ across the load resistance $R$. The power is then $V^2_{rms}/R$. The distortion of the amplifier output can be measured by means of a distortion analyser (see Chapter 14).

**Fault location with analogue circuits**

A service engineer faced with a fault has two basic possible ways of locating it: random testing to find the faulty block with circuits tested in any order, or a systematic approach in which a logical planned sequence of tests is followed. The random approach can occasionally yield a result quicker than the systematic approach, but more often than not will take longer. Systematic approaches can be considered to be based on either reliability data or the functional structure of the unit under test. The reliability approach (see Chapter 4 for a discussion of reliability) involves testing the circuit blocks and components in the sequence of their reliability, the least reliable being tested first. This requires a knowledge of the reliabilities of the blocks and components in the unit. This might, however, be just based on the service engineer's experience of dealing with large numbers of the units and the knowledge that more-often-than-not block X fails first. The most widely used systematic approach to fault finding is, however, based on signal injection and requires a knowledge of the functional structure of the unit, i.e., the functions of the various constituent blocks.

Signal injection involves injecting signals into either inputs or nodes of the circuit and monitoring the signals at the outputs or other nodes. To keep the amount of monitoring to a minimum the procedure called the *half-split technique* is often used when the unit under test is composed of a number of series sub-units. A signal is injected at the input and the output from the unit monitored. If this is in error then the output is monitored of the half-way sub-unit. This then enables the fault to be diagnosed as being the first half or the second half of the unit. The half with the fault in is then tested at its half-way sub-unit. In this way the faulty sub-unit can be determined. For example, with the unit illustrated in Fig. 18.3 the testing sequence might be:

1   With a signal applied to the input, the output at test point 4 is tested. This gives a faulty response.
2   The output at test point 2 is then tested. This gives a correct response. Thus the fault must be in blocks C or D.
3   The output at test point 3 is then tested. This gives a faulty response. Thus the faulty block must be C.

Most systems do not consist of only series connected blocks but also include parallel branches and possibly feedback loops. The term *divergence* is used when the output from one block

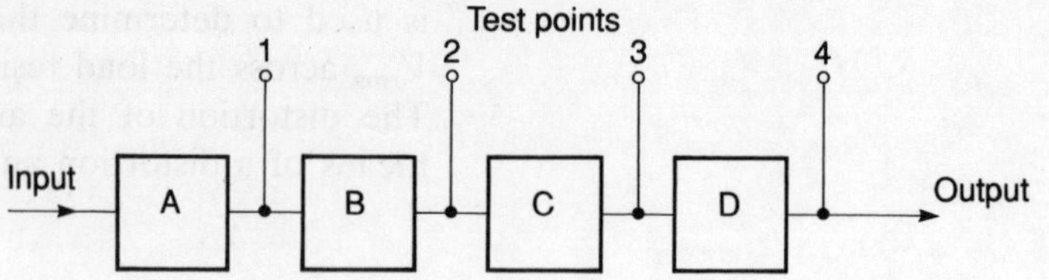

**Test points**

1      2      3      4

Input → A — B — C — D → Output

**Fig. 18.3**   Half-split testing technique

feeds two or more blocks; the term *convergence* when two or more blocks feed a single block. With a divergent system of blocks (Fig. 18.4(*a*)), a fault in block A will affect not only the output from block A but also that from blocks B and C. The diagnosis that block A is faulty will require a tracing back of the signals through the blocks to find the originating failure block. With a convergent system of blocks (Fig. 18.4(*b*)), a fault in block A will lead to the output from blocks A and C being in error. The tracing back of signals through the blocks is necessary to diagnose that block A is the faulty one. If there are feedback loops, i.e., the output of some block is fed back into the input of an earlier block, then there can be difficulties in fault location. The outputs of all the blocks covered by the feedback loop will be in error when the fault is in just one of them or the feedback loop. A knowledge of the system is necessary to determine the best procedure for fault finding; however, one possibility that should be considered is disconnecting the feedback loop at the input point.

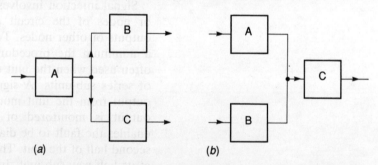

**Fig. 18.4**   (*a*) Divergence,
(*b*) convergence     (*a*)               (*b*)

The basic 'tools' of the test engineer are the maintenance manuals and fault location guides supplied by the manufacturer of the unit under test, and such instruments as a multi-range meter, cathode ray oscilloscope and a signal generator. In some instances more specialist instruments will be required.

**Example 1**

For the audio amplifier system shown in Fig. 18.5, a test engineer carries out the following tests. What is likely to be the faulty unit?

(*a*) Audio frequency signal injected at test point 1 gives a distorted output at 5.

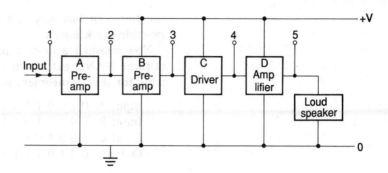

**Fig. 18.5** Example 1

(*b*) Audio frequency signal injected at test point 3 gives normal output at 5.
(*c*) Audio frequency signal injected at test point 2 gives a distorted output at 5.

*Answer*

This is the half-split technique. Test (*a*) indicates that there is a fault in the system. Test (*b*) indicates that the fault is in blocks A or B. Test (*c*) indicates that the fault is in block A.

### Fault location with digital circuits

**Fig. 18.6** Two-input NOR gate

A basic model that is widely used for in-circuit testing with digital circuits is to assume that the effect of any defect is to make one particular node in the circuit remain stuck at either logic state 0 or 1, only one node being directly affected. This fault can, however, have consequential effects on the signals indicated at other nodes. The first step in the devising of a test plan is to list all the possible faults. Thus, for example, in the case of a single NOR gate (Fig. 18.6) we need to consider the possibility of each input being stuck at 0 and 1. A correctly operating NOR gate gives an output of logic 1 when any of its inputs is logic 0. The fault list for a two-input NOR gate is thus

Input A stuck at 0
Input A stuck at 1
Input B stuck at 0
Input B stuck at 1

The next step is to devise tests that could detect each fault. Thus, how can we detect that input A is stuck at 0? What we require is a test which will distinguish between a fault-free input to A and the faulty input. The vital test must thus be one that determines whether input A is at logic 1, since with the fault it cannot realize this state. There is no point in devising a test to determine whether input A is at logic 0 since this is the stuck state. If we make B logic 1 and A logic 1 then the output would be 0 if the input to A is correctly functioning. Because it is stuck at 0 the output will be 1. The test would thus indicate the fault. In this way tests can be devised for each of the possible stuck inputs.

Now consider a more complex system, e.g., one consisting of three NAND gates with convergence (Fig. 18.7). The truth table for the arrangement is

Input A  0 1 0 0 1 1 0 1
Input B  0 0 1 0 1 0 1 1
Input C  0 0 0 1 0 1 1 1
Output   0 0 1 0 1 1 1 1

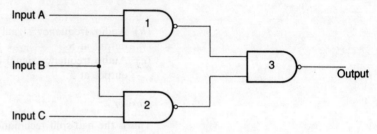

**Fig. 18.7** Convergent system

Consider now a fault which results in input B sticking at 0. A test is required which will determine whether input B is at logic 1. One possibility is to use the input A = 1, B = 1, C = 0 since the sticking input would then give an output of 0 instead of the correct value of 1.

With manual testing of digital systems the basic tools of the test engineer are the logic pulser and the logic tester.

### Example 2

A system of gates (producing an EXCLUSIVE-OR gate system) has the following truth table. What test can be used to determine whether input A is stuck at logic 0?

Input A  0 1 0 1
Input B  0 0 1 1
Output   0 1 1 0

*Answer*

What is required is a test that will distinguish between the outputs given when input A is fault-free and when stuck at 0. Thus if A is made 1 then with B at 0 the correct output will be 1 and the incorrect output 0. Alternatively, if A is made 1 then with B at 1 the correct output will be 0 and the incorrect output 1. Either test could be used.

### Automatic testing

There are three basic forms of ATE:

1  In-circuit component testing where the components in a unit are each tested to find faults, the assumption being made that if each is to specification then so is the unit as a whole.

2  Functional testing where the unit is supplied with signals and measurements made to verify the correct operation of the unit.

3  Combinational testing which combines in-circuit component testing and functional testing.

Combinational testing is more expensive than functional testing and it tends to be more expensive than in-circuit component testing. In-circuit testing requires an individualized bed of nails for each type of board and thus is a more expensive interface than is required with functional testing where edge connectors can be used for the inputs and outputs from the board as a whole. The bed of nails is an array of pins (Fig. 18.8), each of which is located in just the right position to make contact with a circuit node. The pins may be used to inject signals into the circuit or to receive and detect them. In-circuit testing has the advantage of isolating production faults but does not test the performance of the components working together on the board. Functional testing has the advantage of verifying the actual operating specifications of the board but there is difficulty in isolating the fault on the board. Functional testing is faster than in-circuit testing, where every component has to be tested individually.

Whatever the form of testing, whether manual or ATE, it is necessary to work to a test plan to avoid the possibility of missing a fault and also in wasting time, and money, in over-testing.

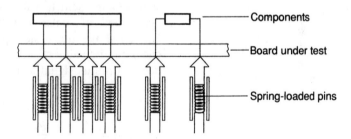

**Fig. 18.8**  Bed of nails

**In-circuit test plans**

The generation of a test plan for in-circuit component testing with, say, a printed circuit board (PCB) requires a consideration of all the possible types of defect that can occur. Thus there can be:

1  Fabrication defects in the integrated circuits or defects in components before insertion into boards.

2  Production defects on the PCB at the bare board stage such as broken or incorrectly routed tracks, missing or extra connections.

3  Production defects on the PCB at the populated stage as a result of the assembly, such as the wrong component, a missing component, component wrong way round, pins not correctly aligned with holes, and due to the soldering, e.g., dry joints or solder splashes joining tracks.

4  Operational defects due to service conditions resulting in components failing as a result of perhaps mechanical shock, electrostatic shock or power dissipation.

The aim of the test plan is to ensure that any PCB with any defect will fail at least one of the applied tests.

Testing on individual components prior to insertion into boards involves determining the characteristics of each device for comparison with those in the specification. Bare board testing involves checking the continuity of the copper track on the printed circuit board to determine whether there are breaks in tracks or short-circuits between tracks. This can be done by using a bed of nails to make contact with a large number of test points on the board, the resistance between test points being checked for continuity and short circuits.

In-circuit testing with digital circuits involves the preparation of fault lists for stuck gates and then the devising of tests that would enable a stuck gate to be identified (see discussion earlier in this chapter). Thus test patterns are generated that can be applied through a bed of nails to circuit nodes, and then the detection of an output departure from the correct value used to indicate which gate is stuck.

In-circuit testing with analogue circuits involves stimulus signals being sent to various test points in the unit under test, via a bed of nails, and the responses determined. Figure 18.9(a) shows the basic type of circuit that might be used for checking the value of a resistance. The resistance under test $R_x$ has a stimulus signal, a d.c. voltage $V_s$, applied to it and the output of the connected operational amplifier monitored. The gain of the inverting operational amplifier is

$$\text{gain} = \frac{V_o}{V_s} = -\frac{R_f}{R_x}$$

Thus

$$R_x = -\frac{R_f V_s}{V_o} \tag{1}$$

Hence $R_x$ can be determined. The above discussion has assumed that the resistor is isolated from the rest of the circuit. Since this is generally not the case, it is necessary to isolate the resistor from the rest of the components on the board. This is done using the adapted circuit shown in Fig. 18.9(b). $R_1$ and $R_2$ represent the resistances of other circuit

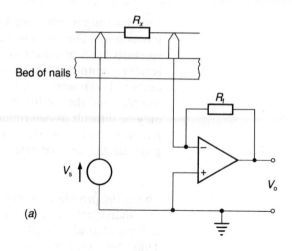

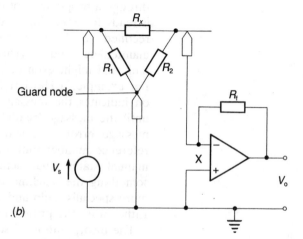

**Fig. 18.9** In-circuit measurement of resistance, (a) basic circuit, (b) guarded circuit

board components. All these other resistances are connected to ground through nails. Point X is a virtual earth, thus there will be no current flow through $R_2$ and so $R_1$ and $R_2$ will not siphon off any of the current that would otherwise flow through $R_x$. Thus $R_x$ is effectively isolated.

**Functional test plans**

Functional testing involves applying test signals to, say, a PCB and determining if the circuit as a whole is correctly working. Thus if the circuit being tested is just a three-input NOR gate then the truth table for that gate indicates

Input A 0 1 0 0 1 1 0 1
Input B 0 0 1 0 1 0 1 1
Input C 0 0 0 1 0 1 1 1
Output  1 1 1 1 1 1 1 0

There are thus eight test patterns of input that can be applied and the output monitored to check if the gate is correctly functioning. The process of testing thus consists of applying successive sets of values to the circuit inputs and observing the resulting outputs, e.g., inputs $A = 0$, $B = 0$, $C = 0$ to give output 1. However, most circuits are considerably more complex and the truth table for the circuit as a whole may not only be difficult to determine but involve a vast number of test patterns. For this reason, computers are often used to generate the test patterns.

**Self-testing**

This is the provision of built-in testing facilities within a unit so that individual parts of the unit will test themselves and can indicate that all is well or that if there is a fault, its location. Thus, for example, when the unit is switched on it may run through a programme of internal tests before indicating that it is ready for operation or that it can not be used until a fault is rectified. Such a system can considerably simplify field maintenance and servicing in that the system identifies the fault and might even indicate the action needed to rectify it. For example, a Hewlett Packard laser printer when switched on indicates the message 'self test' and cannot be operated until the message 'ready' is obtained. If there is a fault, the message error with some reference number is given. The reference number indicates the type of fault and the printer manual states what action is to be taken. This might be something the machine operator can do or it might require more specialist help and the calling out of a service engineer. Either way, the printer has diagnosed what the fault is.

The incorporation of self-testing into a unit does, however, bring with it significant extra costs to the manufacturer. There are additional hardware costs in that extra circuitry and components are needed, increased manufacturing costs because the system is larger, increased design costs, and a decrease in reliability because the more components there are in a system, the smaller the mean time between failures (see Chapter 4). Against this have to be set the reduction in manual or automatic testing costs, and maintenance and servicing costs.

**Problems**

1 At present company X carries out testing only on the finished product and has a 15% rejection rate. Repairs at this stage are costly. The management has for many years considered that testing at earlier stages in the production would be expensive. Present the arguments that might indicate that it is more economic to test at earlier stages.

2 For some unit, e.g., a power supply, devise a manual test specification and specify the test instruments required. You will

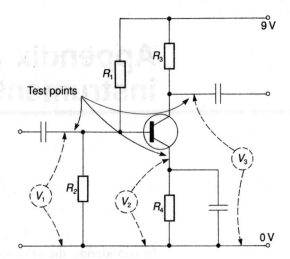

9 V

$R_3$

$R_1$

Test points

$V_3$

$V_1$

$R_2$

$V_2$

$R_4$

0 V

**Fig. 18.10**   Problem 4

need to know how the unit operates and its various constituent elements.

3   Specify measurement methods that can be used in the testing of an amplifier to determine (*a*) the gain, (*b*) distortion, (*c*) input impedance, and (*d*) frequency response.

4   For the single stage transistor amplifier shown in Fig. 18.10, devise a test plan, based on measurements at the three marked test points of the voltages between them and ground, for faults in each resistor resulting in open circuits. You may like to extend this problem into a consideration of the effects of all possible failures, checking the resulting test plan in the laboratory.

5   Explain the difference between in-circuit testing and functional testing and list the advantages and disadvantages of each.

6   Devise a test for a three-input NOR gate that can detect whether one input is stuck at logic 0.

7   Devise a test for a three-input NAND gate that can detect whether one input is stuck at logic 1.

8   For some electronic unit, such as a laser printer, present the arguments for and against including self-testing.

# Appendix A: Criteria for instrument selection

In considering the selection of an instrument, the following can be considered to be a checklist of some of the questions that are likely to need to be answered.

**The quantity being measured**

1 What is the variable that is to be measured, e.g., current?
2 What are the basic characteristics of the quantity that could affect the choice of instrument, e.g., input impedance, loading effects, high frequency, non-sinusoidal, non-periodic rapidly changing current, etc.?

**The environment**

3 What are the environmental conditions that could affect the choice of instrument, e.g., range of temperature, humidity, mechanical shock, vibration, etc.?
4 Is the instrument to be permanently installed in some situation and, if so, what are the size limitations and mounting requirements?
5 If the instrument is to be portable, are there size and weight limitations?
6 What power supplies will be required?

**Accuracy**

7 What is the accuracy required in the measurement?
8 Is the same accuracy required over the entire range of measurement?

**Resolution**

9 What is the smallest change in the quantity being measured that is required to produce an observable change in the instrument reading, i.e., the resolution required?

| | |
|---|---|
| **Range and scale** | 10 What are the maximum and minimum values to be measured?<br>11 Can the range be covered by a single range instrument or is a multi-scale instrument required?<br>12 Is a linear scale required? |
| **Output** | 13 What form of display is required? Is the display to be at some distance from the point of measurement?<br>14 Does the output have to interface with some other system? If so, what signal levels and codes are required? Is there a standard bus? |
| **Response characteristics** | 15 What response time is required?<br>16 What bandwidth is required?<br>17 For a.c. instruments, to what aspect of the waveform should the instrument respond, i.e., peak, mean, r.m.s.? |
| **Calibration** | 18 Is high stability required with the instrument maintaining its calibration over long periods of time?<br>19 What length of time between calibrations is required?<br>20 Is a built-in calibration system required?<br>21 What are the requirements for calibration? |
| **Interference and noise** | 22 Is the quantity being measured floating or has it one side earthed?<br>23 Are there likely to be stray electromagnetic or electrostatic fields?<br>24 What are the required common mode and normal mode rejection ratios? |
| **Reliability** | 25 What is the required reliability?<br>26 Does the instrument include any overload limiting items or alarms?<br>27 What will be the consequences of failure and will a standby instrument be required?<br>28 What are the maintenance requirements and will any special equipment be required?<br>29 Will stocks of special spares need to be kept? |
| **Cost** | 30 Are there any cost limitations on the choice of instrument? |

# Appendix B: Glossary of terms

The following are some of the terms commonly used in specifying the performance of instruments and measurement systems.

*Accuracy* The accuracy of an instrument is the extent to which the reading it gives might be wrong.

*Amplitude* The amplitude of a pulse is measured from the level at which the pulse starts, i.e., the base line, to the steady state pulse value.

*Aperiodic* An instrument is said to be aperiodic when the motion of its index is critically damped or over-damped. See damping.

*Arithmetic mean* This is the sum of all the results $x_1$, $x_2$, etc. divided by the number of results $n$ considered.

$$\text{arithmetic mean} = \frac{x_1 + x_2 + x_3 + \ldots + x_n}{n}$$

*Availability* This is fraction of the total time for which an item is in usable state. It is given by

$$\text{availability} = \frac{MTBF}{MTBF + MTTR}$$

See *mean time between failures* and *mean time to repair*.

*Bandwidth* The bandwidth can be defined as the range of frequencies for which the transfer function is no less than 70.7% of its peak value $G$, i.e., $G/\sqrt{2}$. An alternative way of expressing this is that the bandwidth is the range of frequencies for which the transfer function is within $3\,\text{dB}$ (decibels) of its peak value. A change of $3\,\text{dB}$ means a transfer function which changes by $1/\sqrt{2}$.

*Bel* See *decibel*.

*Bias*  The bias of an instrument is the constant error that exists for the full range of its measurements.

*Binary word*  A word is a grouping of a number of bits. With conventional binary numbers the position of the bits in a word has the significance that the least significant bit (LSB) is on the right end of the word and the most significant bit (MSB) on the left end. The denary value of the bits in a word is

$$2^{n-1} \ldots 2^4\, 2^3\, 2^2\, 2^1\, 2^0$$
MSB                LSB

*Bit*  This is the abbreviation for a binary digit, 0 or 1.

*Calibration*  This is the process of determining the relationship between the value of the quantity being measured and the corresponding position of the instrument index.

*Common mode noise*  This is the noise occurring between the earth terminal of a measurement system and its lower potential terminal.

*Common mode rejection ratio* (CMRR)  This is given by

$$\text{CMRR} = 20\lg(V_{cm}/V_e)$$

where $V_{cm}$ is the peak value of the common mode noise and $V_e$ the peak value of the error it produces at a particular frequency.

*Conversion time*  This is the time it takes an analogue-to-digital converter to generate a complete digital word when supplied with an analogue input, or vice versa for a digital-to-analogue converter.

*Cross talk*  This is the interference that occurs between neighbouring channels with a multiplexer or other element which has parallel inputs.

*Damping*  An instrument is said to be damped when the amplitude of the free oscillations of the index are progressively reduced or completely suppressed. The instrument is under-damped when oscillations occur, critically damped when the degree of damping is just sufficient to prevent oscillation and over-damped when the degree of damping is more than sufficient to prevent oscillation. See *aperiodic*.

*Dead space*  The dead space of an instrument is the range of values of the quantity being measured for which it gives no reading.

*Decibel*  The ratio between two values of electric or acoustic power is usually expressed on a logarithmic scale. With base-10 logarithms the ratio is given the unit the bel. The decibel is one-tenth of a bel.

$$N_{bel} = \lg(P_1/P_2)$$

$$N_{dB} = 10\lg(P_1/P_2)$$

*Deviation*  The deviation is the amount by which an individual measurement departs from the mean. Thus for a measurement of $x_1$

$$\text{deviation} = x_1 - \bar{x}$$

with $\bar{x}$ being the mean value.

*Discrimination*  The discrimination of an instrument is the smallest change in the quantity being measured that will produce an observable change in the reading of the instrument.

*Dissipation factor*  This is the reciprocal of the Q factor and is generally used to describe the quality of capacitors. See *Q-factor*.

*Drift*  An instrument is said to show drift if there is a gradual change in output over a period of time which is unrelated to any change in input. See *zero drift*.

*Droop*  This is the amount by which the peak value of a pulse decreases during the pulse.

*Duty cycle*  This term is used to define the ratio of the pulse width to pulse cycle time or period.

$$\% \text{ duty cycle} = \frac{\text{pulse width}}{\text{pulse period}} \times 100\%$$

*Dynamic range*  This is the range of signals between the smallest that can be detected above the noise of the system and the largest signal that does not cause any spurious signals greater than the smallest signal that can be detected.

*Effective range*  The effective range is that part of the scale over which measurements can be made to the specified accuracy.

*Error*  The error of a measurement is the difference between the result of the measurement and the true value of the quantity being measured.

$$\text{Error} = \text{measured value} - \text{true value}$$

*Failure rate*  This is the average number of failures per item per unit time. Thus if $N$ items are tested for a time $t$ with failed items being repaired and put back into service, then if during that time there are $N_f$ failures

$$\text{failure rate} = \frac{N_f}{Nt}$$

*Fiducial value*   This is the quantity to which reference is made in order to specify the accuracy of an instrument. It is usually the full-scale reading.

*Form factor*   This is

$$\text{form factor} = \frac{\text{r.m.s. value of a periodic waveform}}{\text{mean value of half a cycle}}$$

*Gain*   The gain of a system or element is the size of the output divided by the size of the input:

$$\text{gain} = \frac{\text{output}}{\text{input}}$$

*Hysteresis*   Instruments can give different readings, and hence an error, for the same value of measured quantity according to whether that value has been reached by a continuously increasing change or a continuously decreasing change. This effect is called hysteresis. The hysteresis error is the difference between the measured values obtained when the measured quantity is increasing and when decreasing to that value. Hysteresis is often expressed in terms of the maximum hysteresis as a percentage of the full-scale deflection.

*Indicating instrument*   This is a measuring instrument in which the value of the measured quantity is visually indicated, e.g., a pointer position on a scale.

*Interference*   This is noise due to the interaction between external electrical and magnetic fields and the measurement system circuits.

*Lag*   A system is said to show lag if, when the quantity being measured changes, the measurement system does not respond instantaneously but some time later.

*Limiting error*   In the case of some components and instruments their deviations from the specified values are guaranteed to be within a certain percentage of those values. The deviations in this case are then referred to as limiting errors or tolerances.

*Mean*   See *arithmetic mean*.

*Mean deviation*   This is the mean of the deviations, ignoring the signs of each deviation.

$$\text{mean deviation} = \frac{|x_1 - \bar{x}| + |x_2 - \bar{x}| + \ldots + |x_n - \bar{x}|}{n}$$

*Mean time between failures* (MTBF)   If $N$ systems are tested for a time $t$ with failed items being repaired and put back into service, then if during that time there are $N_f$ failures

$$\text{MTBF} = \frac{Nt}{N_f}$$

*Mean time to failure* (MTTF)   The mean time to failure is the average time to failure for a number of samples of a product, it being assumed that it is impossible or uneconomic to repair them. Thus if $N$ items are tested and the time to failure for each is $t_1$, $t_2$, $t_3$, . . ., $t_N$, then the average time to failure is

$$\text{MTTF} = \frac{t_1 + t_2 + t_3 + \ldots + t_N}{N}$$

*Mean time to repair* (MTTR)   This is the average time it takes to repair a system.

*Noise*   This is used to describe the unwanted signals that may be picked up by the measurement system and interfere with the signal being measured, thus giving rise to random errors.

*Noise figure* (NF)   This is a measure of the amount of noise introduced by using an instrument and is given by

$$\text{NF} = 10\lg\left[\frac{(\text{S/N})_{\text{in}}}{(\text{S/N})_{\text{out}}}\right]$$

where $(\text{S/N})_{\text{in}}$ is the signal-to-noise ratio of the input and $(\text{S/N})_{\text{out}}$ that of the output.

*Non-linearity error*   A linear relationship for an element or a system means the output is directly proportional to the input. In many instances, however, though a linear relationship is used it is not perfectly linear and so errors occur. The non-linearity error is the difference between the true value and what is indicated when a linear relationship is assumed. Non-linearity is often expressed in terms of the maximum non-linearity error as a percentage of the full-scale deflection.

*Normal mode noise*   This is all the noise occurring within the signal source that is being measured.

*Normal mode rejection ratio* (NMRR)   This is given by

$$\text{NMRR} = 20\lg(V_n/V_e)$$

where $V_n$ = peak value of normal mode noise, $V_e$ = peak value of the error it produces at a particular frequency.

*Offset*   The term offset is used to describe the deviation of the output signal of a system from zero when the input is zero.

*Operating temperature*   This is the temperature range within which the instrument can be used and give the stated accuracy, stability, etc.

*Output impedance*   This is the impedance between the output terminals of the instrument.

*Overshoot*  This is the deviation from the peak value of a pulse immediately following a rising edge. Alternatively, it can be used to describe the amount by which an instrument pointer overshoots the steady state value when there is a sudden input to the instrument.

*Persistence time*  This term is used with cathode ray tubes to define the time it takes for the screen phosphor to reach decay to 10% of its luminance when electrons cease bombarding it.

*Precision*  This is a measure of the scatter of results obtained from measurements as a result of random errors. It describes the closeness of the agreement occurring between the results obtained for a quantity when it is measured several times under the same conditions.

*Preshoot*  The preshoot of a pulse is the deviation that occurs in the base line at the start of the pulse.

*Primary standards*  These are the fundamental basic unit standards maintained by a country's national standards laboratory.·

*Probable error*  A statement of the probable error for a set of measurements means that there is a 50% chance that if we take just one measurement then it will have a random error no greater than $\pm 0.6745\sigma$ from the mean value, where $\sigma$ is the standard deviation.

*Pulse fall time*  This is the time required for a pulse to fall from 90% to 10% of its normal amplitude.

*Pulse repetition rate*  This is the frequency with which a pulse occurs and is equal to the reciprocal of the pulse cycle time or period.

*Pulse rise time*  This is the time required for a pulse to rise from 10% to 90% of its normal amplitude.

*Pulse width*  This is the time the pulse takes from the 50% amplitude point on the leading edge to the 50% point on the trailing edge.

*Q-factor*  This is defined as

$$Q = \frac{2\pi \times \text{maximum energy stored in a cycle}}{\text{energy dissipated per cycle}}$$

This can be shown to be, for a capacitor considered as capacitance in series with resistance, or an inductor as inductance in series with resistance,

$$Q = \frac{\text{reactance}}{\text{resistance}}$$

*Quantization error* This is the error with a digital-to-analogue or analogue-to-digital converter due to the quantization interval, this being the contribution of the least significant bit.

*Quantization noise* This is the noise that can be considered to be added to an analogue signal as a consequence of the quantization error.

*Random errors* These are errors which vary in an unpredictable manner between successive readings of the same quantity, varying both in the size of the error and whether the error is positive or negative.

*Random noise* This is noise which is due to the random motion of electrons and other charge carriers in components and is a characteristic of the basic physical properties of components in the system.

*Range* The range of an instrument is defined as the limits between which readings can be made.

*Reliability* The reliability of a product is defined as the chance that it will operate to a specified level of performance for a specified period under specified environmental conditions.

*Repeatability* The repeatability of an instrument is its ability to display the same reading for repeated applications of the same value of the quantity being measured.

*Reproducibility* The reproducibility of an instrument is its ability to display the same reading when it is used to measure a constant quantity over a period of time or when that quantity is measured on a number of occasions.

*Recording instrument* This is a measuring instrument in which the values of the measured quantity are recorded.

*Resolution* The resolution of an instrument is the smallest change in the quantity being measured that will produce an observable change in the reading of the instrument. When used with an analogue-to-digital converter it represents the smallest change in analogue input that will generate a change of one bit.

*Response time* When the quantity being measured changes, a certain time, called the response time, has to elapse before the measuring instrument responds fully to the change.

*Ringing* This is the oscillation that occurs in the height of a pulse immediately following a rising edge.

*Rise time* This term is used with cathode ray oscilloscopes for the time taken for the deflection to go from 10% to 90% of its steady deflection when there is a step input.

*Root-mean-square values* For a periodic waveform the root-mean-square current or voltage is the square root of the sum of the mean values of the squares of the currents or voltages. For a sinusoidal waveform it is the maximum value divided by $\sqrt{2}$.

*Sag* is the amount by which the peak value of a pulse decreases during the pulse.

*Sample rate* Some instruments take samples of the variable at regular intervals. The greater the sample rate, i.e., the greater the number of samples taken per second, the more readily the instrument readings mirror a rapidly changing input.

*Scale* This is the array of marks, together with associated figuring, against which the position of a pointer, light spot, liquid surface or other form of index is indicated.

*Scale factor* This is the factor by which the indicator readings have to be multiplied to obtain the value for the measured quantity.

*Scale interval* This is the amount by which the measured quantity changes in the index moving between adjacent scale marks.

*Scale length* This is the distance, measured along a line that defines the path of the index, between the end marks of the scale.

*Secondary standards* These are reference standards which are periodically verified and calibrated against the national laboratory's primary standard (see *primary standards*).

*Sensitivity* The sensitivity of an instrument is

$$\text{sensitivity} = \frac{\text{change in instrument scale reading}}{\text{change in the quantity being measured}}$$

*Sensitivity drift* The sensitivity drift is the amount by which the sensitivity changes as a result of changes in environmental conditions.

*Settling time* The settling time is the time required for the ringing of a pulse to decrease to a given percentage, usually 1–5%, of the overshoot. See *ringing* and *overshoot*.

*Signal-to-noise ratio* This is the ratio of the signal level $V_s$ to the internally generated noise level $V_n$. It is usually expressed in decibels, i.e.,

$$\text{signal-to-noise ratio in dB} = 20 \lg (V_s/V_n)$$

*Slew rate* The slew rate is the maximum rate of change with time of the input for which the output can keep up with the change.

*Span* The span of an instrument represents the limits between which readings can be made.

*Stability* The stability of an instrument is its ability to display the same reading when it is used to measure a constant quantity over a period of time or when that quantity is measured on a number of occasions.

*Standard* This, whether it be a material standard of a specification of a test method, provides the method of comparison by which an instrument is calibrated.

*Standard deviation* The deviation for a measurement is the difference between the mean and the value of that measurement. The sum of the squared deviations for all the measurements obtained divided by the number of measurements $n$ gives the mean of the squares of the deviations. The square root of this quantity is the standard deviation $\sigma$.

$$\sigma = \sqrt{\left[ \frac{(x_1 - \bar{x})^2 + (x_2 - \bar{x})^2 + \ldots + (x_n - \bar{x})^2}{n} \right]}$$

*Systematic errors* These are errors that remain constant with repeated measurements.

*Testability* A testable design is defined as being one that has built-in facilities that allow simple, efficient and effective testing to be carried out.

*Threshold* This is the minimum value a signal must have reached before the instrument responds and gives a detectable reading.

*Tolerance* See *limiting error*.

*Traceability* All standards should be calibrated against superior standards that are traceable to national or international standards.

*True value* This is the value with zero error.

*Undershoot* The undershoot of a pulse is the deviation that occurs in the base line immediately following the end of the pulse.

*Unreliability* The unreliability of a system is the chance that the system will fail to operate to specification for a specified period of time under specified environmental conditions.

Unreliability = 1 − reliability

*Word* See *binary word*.

*Working standards*  These are day-to-day standards used to check and calibrate instruments or to carry out comparison measurements. They are calibrated against secondary standards (see *secondary standards*).

*Writing time*  This term is used with cathode ray tubes to define the time taken from the beginning of the excitation of the phosphor to it reaching 90% of the maximum intensity plus the persistence time (see *persistence time*).

*Zero drift*  This term is used to describe the change in the zero reading of an instrument that can occur with time.

# Answers to problems

**Chapter 1**

1. See text
2. (*a*) $-5\,$mA, (*b*) $-4\%$
3. See text, accurate = small error, precision = small random error
4. (*a*) $10.08\,\Omega$, (*b*) $0.068\,\Omega$, (*c*) $0.078\,\Omega$
5. (*a*) $9.0\,$V, (*b*) $0.83\,$V, (*c*) $1.2\,$V
6. $2.85\,$kV, $0.25\,$kV
7. (*a*) $101.3\,\Omega$, (*b*) $0.20\,\Omega$, (*c*) $0.13\,\Omega$
8. $\pm 3.0\,\Omega$
9. (*a*) $9.63\,\mu$F, (*b*) $0.21\,\mu$F, (*c*) $0.14\,\mu$F
10. $0.500 \pm 0.075\,$W
11. (*a*) $150 \pm 12.5\,\Omega$, (*b*) $33.3 \pm 7.8\,\Omega$
12. $400.0 \pm 4.8\,$W
13. $240\,\Omega \pm 2.5\% = 240 \pm 6\,\Omega$
14. $y = 1.6x + 4.0$
15. $y = 2.2x - 1.5$
16. $0.71$, $0.46$
17. $y = 3.31x - 2.6$, $x = 0.30y + 0.84$, $0.81$
18. $y = 2.95x - 3.47$, $x = 0.32y + 1.33$, $0.97$
19. (*a*) log–log, (*b*) linear--linear, (*c*) log–linear
20. $\ln R_t = \beta/t + \ln a$, log–linear
21. $\lg l = n\lg T + \lg k$, log–log graph, $n = 2$, $k = 0.25$
22. (*a*) $R$ against $1/d^2$ or $\lg R$ against $\lg d$, (*b*) $R$ against $L$.
23. (*a*) Line graph, (*b*) pie chart, (*c*) histogram, (*d*) bar chart, (*e*) histogram, (*f*) bar chart

**Chapter 2**

1. (*a*) $\pm 30\,$Hz, (*b*) $\pm 30\,$kHz
2. $5.0 \pm 0.2\,$ms
3. (*a*) $\pm 3$, (*b*) $\pm 1.7$
4. $0.15\,°$C
5. $200\,\mu$V
6. Takes $1\,$s before the meter fully responds to a full-scale current/voltage

7. The input has to be greater than 0.3% of the span before it responds
8. For the same input on different occasions the reading may vary by 0.05%
9. The minimum detectable change is $1\,\mu V$
10. 7.0 dB
11. (a) 10, (b) 0.32
12. $12 \pm 0.06\,V$
13. See text

**Chapter 3**

1. (a) 0.10%, (b) 4.8 %
2. 14.3 $\Omega$
3. 3%
4. 4.8%
5. 0.13 V
6. (a) 5.7 $\mu V$, (b) 9.0 $\mu V$, (c) 10.6 $\mu V$
7. See text. $\sqrt{(298/273)} = 1.04$
8. 80 dB
9. 80 dB
10. See text
11. See text. Capacitative coupling
12. See text

**Chapter 4**

1. 0.9996 or 99.96%
2. 0.10
3. (a) 0.004 per hour, (b) 250 hours
4. $2.5 \times 10^7$ hours
5. 0.998
6. (a) 0.99, (b) 0.61, (c) 0.37, (d) 0.0067
7. 0.98
8. 0.975
9. 0.44
10. 0.93
11. 0.9999
12. 0.9999

**Chapter 5**

1. See text
2. See text; consider realization world-wide and ability to be stored
3. See text and Fig. 5.4

**Chapter 6**

1. All deflections halved
2. (a) 0.25 $\Omega$, (b) 0.010 $\Omega$, (c) 950 $\Omega$, (d) 19 950 k$\Omega$
3. 0.00505 $\Omega$, 0.04545 $\Omega$, 0.4495 $\Omega$

4. $0.050\,\Omega$, $0.453\,\Omega$, $4.522\,\Omega$
5. Total multiplier resistance for each range $950\,\Omega$, $9950\,\Omega$, $99\,950\,\Omega$
6. $10\,k\Omega/V$
7. $-2.9\%$
8. $33\%$
9. $492\,\Omega$, $10\,\Omega$
10. $I_{av} = 0.64I_m$, $I_m$, $0.70I_m$
11. $5.0\,A$, $7.1\,A$
12. See text
13. Scale converted to r.m.s. values assuming sinusoidal.
14. Suggestions might be ($a$) thermocouple meter, ($b$) moving iron or hot wire meter, ($c$) moving coil meter, ($d$) electronic meter.

**Chapter 7**

1. See text
2. $24.4\,mV$
3. $0.098\%$
4. See text
5. $500$
6. $8\,\mu s$
7. ($a$) Ramp, ($b$) successive approximations or flash depending on how fast, ($c$) an integrating ADC, e.g., dual ramp
8. 4-digit display with the MSD only able to be 0 or 1
9. $0.1\,V$

**Chapter 8**

1. See text
2. E.g. rectilinear chart for ($b$) rather than curvilinear in ($a$); greater sensitivity and bandwidth for ($c$)
3. $0.5$, $\times 0.73$
4. $60\,\Omega$ in series
5. ($a$) $0.22\,rad/V$, ($b$) $2.0$, ($c$) $3.6\,Hz$
6. E.g. potentiometric type have higher input resistance, higher accuracy but slower reading and only usable for d.c.
7. See text
8. See text

**Chapter 9**

1. $\pm 2\%$ or more likely $\pm 1.4\%$
2. To minimize loading (see Chapter 3)
3. Accuracy of meters, loading
4. $8.0\,\Omega$
5. $0.044\,V$

6. $6.1\,\mu A$
7. $16.7\,mm$
8. See text and Fig. 9.7
9. $0.23\,V$
10. $0.021\,\Omega$
11. $400\angle 80°\,\Omega$
12. $20\,\Omega$, $20\,mH$
13. $49\,mH$, $18.4\,\Omega$
14. $2.4\,\mu F$, $0.83\,k\Omega$
15. $0.20\,\mu F$, $55\,\Omega$, $0.069$
16. $0.050\,\mu F$, $860\,\Omega$, $0.043$
17. $9.6\,mH$, $20\,\Omega$
18. See text
19. See text and Fig. 9.31
20. $7.3\,pF$
21. $160\,pF$, $3.5\,k\Omega$

**Chapter 10**

1. See text and Fig. 10.2
2. See text and Figs 10.3 and 10.4
3. $50\,W$, a linear scale
4. $10\,kW$, $0.82$
5. See text and Fig. 10.12
6. Sampling wattmeter, time to take samples

**Chapter 11**

1. See text and Fig. 11.1
2. High luminance, short persistence, as with P31
3. $35\,ns$
4. See text
5. $1\,M\Omega$, $44\,pF$
6. Under compensation, see text
7. (a) Storage oscilloscope, (b) dual beam or dual trace oscilloscope, (c) storage oscilloscope, (d) a basic real-time oscilloscope, (e) a storage oscilloscope
8. See text; different waveforms fit same sample points and hence need for high frequency sampling ADC
9. See text
10. $80\,V$, $133\,Hz$
11. 5 to 1
12. See text

**Chapter 12**

1. $23.54\,kHz$
2. 1 in 1000
3. See the text on gating error
4. See text
5. 1 part in 999 999; $9.999\,99\,MHz$

6. ±0.000 35%
7. ±0.1%
8. See text

**Chapter 13**

1. Hartley and Colpitts 10 kHz to 100 MHz, Wien up to 1 MHz, phase shift up to 10 MHz
2. See text
3. 25%
4. See text and Fig. 13.3
5. See text and equation [3]
6. See text and Fig. 13.4
7. See text and Figs 13.7 and 13.8; direct uses harmonics, indirect a divider with a phase-locked loop.

**Chapter 14**

1. See text and Fig. 14.4
2. See text
3. (a) 10 MHz, 9.995 MHz, 10.005 MHz (b) 0.06
4. (a) 1 MHz, 0.995 MHz, 1.005 MHz, (b) 0.63
5. (a) 0.8%, (b) 1.7%
6. See text

**Chapter 15**

1. See text
2. B
3. See text and Fig. 15.5
4. See text
5. See text
6. See text

**Chapter 16**

1. See text and Fig. 16.4; serial – cheap but slow, parallel – quick but expensive
2. See text
3. See text; data valid, not ready for data, not data accepted
4. See text, simplification of interconnection
5. See text, start and stop bits
6. See text and Fig. 16.2

**Chapter 17**

1. E.g. (a) thermocouple, (b) photo-conductive cell, (c) resistance strain gauge, (d) potentiometer.
2. See text
3. 0.059 V
4. $5.25 \times 10^{-5}$ V
5. $2.49 \times 10^{-8}$ F
6. 0.05 mm

7. Your design might use four cells to support the tank, with a Wheatstone bridge giving an out-of-balance reading
8. The volume of float immersed determines the force exerted and the form of the cantilever the amount of strain experienced
9. See text
10. See text, maximum rate of change of input signal that can be resolved
11. See text, e.g. parallel/serial transmission, distances, rate of transmission, resistance to external noise, number of multiplexed signals
12. E.g. thermistors, conversion to 4–20 mA current, analogue transmission, multiplexer, sample-and-hold, ADC, microprocessor, display

**Chapter 18**

1. See text
2. Consider continuity, insulation and operation with power on
3. See text
4. $R_1 = 0$, $V_1 = 9\,\text{V}$, $V_2 = V_3 = 0$; $R_2 = 0$, $V_1$ below normal, $V_2$ and $V_3$ above normal; $R_3 = 0$, all signals below normal. With $R_1 = 0$ or $R_3 = 0$ there is no output, with $R_3 = 0$ the output is distorted
5. See text
6. A stuck at 0, apply A = 1, B = C = 1 when error output = 1
7. A stuck at 1, apply A = 0, B = C = 0 when error output = 0
8. See text

# Index